THE FUTURE OF THE PAST

Built Heritage Research and Practices

ISSN: 2767-5343
eISSN: 2767-5351

Volume 1

PROCEEDINGS OF THE FUTURE OF THE PAST: PARTICIPATORY GOVERNANCE, CUENCA, ECUADOR, 2–6 DECEMBER 2020

The Future of the Past: Paths towards Participatory Governance for Cultural Heritage

Edited by

Gabriela García

City Preservation Management project, Facultad de Arquitectura y Urbanismo, Universidad de Cuenca, Ecuador

Aziliz Vandesande

Raymond Lemaire International Centre for Conservation, Department of Civil Engineering, KU Leuven, Heverlee, Belgium

Fausto Cardoso

City Preservation Management project, Facultad de Arquitectura y Urbanismo, Universidad de Cuenca, Ecuador

Koen Van Balen

Raymond Lemaire International Centre for Conservation, Department of Civil Engineering, KU Leuven, Heverlee, Belgium

CRC Press
Taylor & Francis Group
Boca Raton   London   New York   Leiden

CRC Press is an imprint of the
Taylor & Francis Group, an **informa** business

A BALKEMA BOOK

CRC Press/Balkema is an imprint of the Taylor & Francis Group, an informa business

© 2021 Taylor & Francis Group, London, UK

Typeset by MPS Limited, Chennai, India

Library of Congress Cataloging-in-Publication Data
Applied for

Published by: CRC Press/Balkema
                Schipholweg 107C, 2316 XC Leiden, The Netherlands
                e-mail: Pub.NL@taylorandfrancis.com
                www.routledge.com – www.taylorandfrancis.com

ISBN: 978-1-032-02129-4 (Hbk)
ISBN: 978-1-032-02130-0 (Pbk)
ISBN: 978-1-003-18201-6 (eBook)
DOI: 10.1201/9781003182016

The Future of the Past:
Paths towards Participatory Governance for Cultural Heritage – García et al (eds)
© 2021 Taylor & Francis Group, London, ISBN 978-1-032-02129-4

Table of contents

The Future of the Past:
Paths towards Participatory Governance for Cultural Heritage – García et al (eds)
© 2021 Taylor & Francis Group, London, ISBN 978-1-032-02129-4

Committees

ORGANISING COMMITTEE

Prof. Fausto Cardoso, *Universidad de Cuenca, Ecuador*
Dr. Gabriela García, *Universidad de Cuenca, Ecuador*

INTERNATIONAL SCIENTIFIC COMMITTEE

Marc Craps, *Belgium*
Ana Pereira Roders, *The Netherlands*
Margherita Sani, *Italy*
Stefano della Torre, *Italy*
Koen Van Balen, *Belgium*
Pieter Van Den Broeck, *Belgium*
Aziliz Vandesande, *Belgium*
Christian Ost, *Belgium*
Eduardo Rojas, *USA*
Monica Lacarreu, *Argentina*
Sebastián Sepúlveda, *Chile*
Diana Burgos-Vigna, *France*
Joaquín Farinós Dasí, *Spain*
Cristina Zurbriggen, *Uruguay*

LOCAL SCIENTIFIC COMMITTEE

Maria Eugenia Siguencia, *Ecuador*
Luis Herrera, *Ecuador*
Dora Arízaga, *Ecuador*
Gabriela López, *Ecuador*
Gabriela Eljuri, *Ecuador*
Xavier Izko, *Ecuador*

Introduction

The Future of the Past:
Paths towards Participatory Governance for Cultural Heritage – García et al (eds)
© 2021 Taylor & Francis Group, London, ISBN 978-1-032-02129-4

Paths towards participatory governance for cultural heritage

G. García
City Preservation Management Project, Faculty of Architecture, University of Cuenca, Cuenca, Ecuador

A. Vandesande
Raymond Lemaire International Centre for Conservation, Department of Civil Engineering, KU Leuven, Heverlee, Belgium

1 INTRODUCTION

El *Futuro del Pasado* or The *Future of the Past*, is an initiative that emerged in Ecuador in 2014, in the frame of the 15th anniversary of the nomination of Santa Ana de los Ríos de Cuenca as World Heritage Site (WHS). It was driven by the research project of the Faculty of Architecture of the University of Cuenca, named "City Preservation Management" (CPM) (Figure 1). At that time, this research project, created in 2007 in close collaboration with the Raymond Lemaire International Centre for Conservation of KU Leuven, had consolidated an academic approach for studying particularities of cultural heritage, adapting methodological reflections, and creating common understandings between academics and researchers.

The theoretical basis, was progressively enriched with the joint work carried out with political and administrative representatives of government institutions, including also religious actors. This joint work (academic and public) revealed a new face of heritage management, including the needs, risks and perspectives from the point of view of policy makers and public representatives.

A large shift in the City Preservation Management approach came in 2011, when the project moved beyond involving representative actors, to design intervention practices to involve citizen's perspectives. Those initiatives were crucial to link theories and practices, and to favor engagement of this strategic actor (citizens) who are still too ignored, in the process of preserving cultural heritage.

Based on these experiences, the CPM project started organizing bi-annual the Future of the Past conference as an academic space of a dialogue to exchange with a variety of local actors and to acknowledge the diverse visions about cultural heritage, including inhabitants, owners and neighbors. These bi-annual local aimed at knowledge exchanges and learning process between diverse local actors, to disseminate initiatives and to promote coordinated actions based on shared responsibility to protect heritage values and pass them on to future generations.

To consolidate and upscale these very practical experiences, the third edition of the Future of the Past conference, was jointly organized by CPM and, the Raymond Lemaire International Centre for Conservation of KU Leuven (Cuenca, Ecuador, 2–6 December, 2019) which met for first time international and local speakers from 10 different countries.

This volume reports on the selected lectures presented from that conferences, which aimed to tackle what remains today one of the major challenges in the cultural heritage field. The different chapters in this book reflect on participatory cultural heritage management experiences and their contribution to consolidate an effective participatory governance process, hence aiming towards: "Paths towards participatory governance of cultural heritage".

2 METHODOLOGY

Similar to the previous local editions, this international edition of the Future of the Past created a dialogue between diverse actors involved in cultural heritage field. However, for the very first time, on this occasion were included insights from international experts. a total of 23 experts from ten different countries (Italy, Belgium, France, Spain, The Netherlands, USA, Argentina, Uruguay, Chile and Ecuador) give support to this academic event as part of the scientific committee, while around 120 local participants – representing academics, technicians, representative of

Figure 1.   Logo, City Preservation Management project.

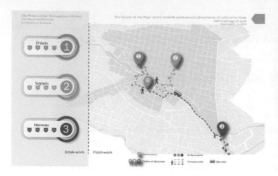

Figure 2. Methodology of work linking desk-work with field-work. The Future of the Past Conference.

Figure 3. Thematic discussion tables, after transect walks. The Future of the Past Conference.

public institutions, civil society, and students – met in Cuenca during four intensive days or lectures, workshops and transect walks. During the opening day of the *paths towards participatory governance of cultural heritage* different reflections on the 20th anniversary of the nomination of the Historic Centre of Cuenca as World Heritage Site were presented. This coincided with the 10th anniversary of the UNESCO Chair Preventive Conservation, Maintenance and Monitoring of Monuments and Sites (PRECOM^{3}OS), established at the KU Leuven. Being aware that a large quantity and diverse actors is not enough to reach real participation, this international conference adopted an innovative strategy of work. Taking advantage of previous experiences, CPM built a platform for exchange between international and local actors, moving the dialogue beyond the mere exchange of ideas to create listening spaces to joint reflection.

In the three following days (December 3–5), the discussion followed the different topics, one per day. The lectures of each morning, at the Auditorium (Aula Magna) of the Faculty of Architecture and Urbanism (FAUC) at Universidad de Cuenca were combined during the afternoon with transect walks through traditional neighborhoods in the historic city: El Vado, Sagrario, and Las Herrerías (Figure 2).

In this manner, a keynote speaker presented an introductory lecture, followed by the presentation of papers related to the topic, and reflections on site. The selected neighborhoods are considered as repositories of significant cultural heritage values (tangible and intangible) which are currently at risk and where conditions of habitability are progressively decreasing. During the transect walks participants interacted with local actors and were invited to respond questions related to the different conference topics. At the end of each transect walk, participants work in groups around thematic discussion tables were the individual inputs collected during the transect walks were presented (Figure 3). Using the world coffee technique (Paño et al. 2019), participants were invited to identify key aspects and to draft recommendations. The results were presented in plenary at the end of each afternoon session and a summary was presented to all participants in the closing ceremony at the end of

the academic event. This volume gathers some of the contributions of selected authors in the frame of the conference.

3 CONTENT AND SCOPE OF THIS VOLUME

The aim of a heritage project functioning as an instrument to engage with disparate groups within society is not new. Participatory approaches, public participation, community involvement, engaging civil society, heritage communities, etc. are increasingly used expressions in heritage management and conservation. Considering the divers spectrum of terms used in this context, community participation can have different natures ranging from information sharing to shared research, from shared heritage valuing to shared heritage management (Vandesande et al. 2018).

Cultural strategies with the objective of achieving greater social equity and more socially inclusive urban environments have been developed since the late 1990s (McLoughlin et al. 2006; Landorf 2011). On the one hand, this development is reflected in the research field, as a clear rise in studies on the social impact of cultural heritage can be found following the 1990s, both within the EU CHCFE Consortium (2015), and beyond (Garcia et al. 2018). However, tracing empirical proof of the effects that cultural heritage can produce at a social level remains a complex exercise, requiring an assessment of impact on an individual scale, as well as an examination of how heritage projects influence the societal level within a longer timeframe.

On the other hand, this development was an important step to our understanding that successful heritage development projects should include a wide variety of actors on the different components of the quintuple helix model. Namely the higher education system (research, education), the economic system (business partners, contractors, consultants), the political system (government, public administration), culture based public system (civil society, NGOs) and finally the actors in the societal or natural environment of built heritage resources (Vandesande & Van Balen 2019). While the more active participation of these actors is highly sought after, the question of ensuring real collaborating in and co-creating project ideas – and of course how to govern these processes – remains on the drawing table of both researchers and policy makers alike.

According to the European Union report drafted by different member state experts through the Open Method of Coordination (2018: 29), "*during the varied discussions about stronger and more active civic participation in public affairs, the focus on cultural policy and at cultural heritage institutions has been changing from governing and participation to a participatory governance approach*".

Thus, Participatory governance emerges as result of the inability of the institutional structures to provide sustainable solutions, and contributes to the improvement of democracy (Vidovic, 2018). Participatory governance is understood as institutional decision-making structures supported by shared responsibilities and rights among diverse actors that goes beyond conventional methods mainly lead by institutional actors. It tries to encompass both scientific and popular discourses, and to be interactive and open to different perspectives and views. Since its theory emerged, progressively has increasingly been accepted and several countries included Ecuador, have put important efforts on creating the legal basis to favor participatory governance.

However, the practical implementation of this approach seems often ambitious and not easy to apply (Vidovic, 2018). Unfortunately, participatory approach has been affected by a growing skepticism. According to Villasante (García et al. 2020), this responds to reiterative misleading consultative practices from the political side, which are aggravated due to the low interest of social actors or sectors of being part of a real participatory process.

According to Vidovic (2018), one of the first documents in the international context that linked participation with cultural heritage, was The Convention for the Safeguarding of Intangible Cultural Heritage, adopted in 2003 at the UNESCO General Conference. Since then, European countries lead the debate to promote participatory approaches to cultural policy-making. This issue was expanded to an international context through instruments such as the Agenda 21 for Culture (UCLG 2004).

Despite these advances and the relevance of participatory governance in culture, it is still not well known how much the guidelines and recommendations of such documents are applied in practice specially in the Latin American context. In this sense, this volume presents an overview about reflections and experiences in the topic that emerge from the two sides – North and South – of the global context. Contents are organized in three main chapters that follow the thematic axes: (1) participatory management of private and public cultural heritage, (2) cultural heritage as a source of inspiration for new participatory management approaches, and (3) lessons from territorial participatory management for effective participatory governance systems in the cultural heritage field.

3.1 *Participatory management of private and public cultural heritage*

The keynote of this chapter is a contribution by **M. Craps**, who has been working as professor at KU Leuven, Faculty of Economics and management, centre for corporate sustainability, member of the sustainability council and interfaculty council for development cooperation at his university, promotor and co-promotor of various vlir-UOS projects. For more than 15 years, his main research interest has been focused on the quality of the relationships between partners in collaborative multi-stakeholder initiatives dealing with complex sustainability issues. In this article, and concerning the cultural heritage field, Craps invites to move towards a collaborative governance model which goes beyond the government structures to give space to a co-creation process with the active participation of private, community and academic actors who adopt new roles and responsibilities.

In line with this theoretical refection to collaborative governance, **K. Van Balen** summarizes the insights gained a long one decade of the UNESCO Chair on Preventive Conservation, Monitoring and Maintenance of Monuments and Sites (PRECOM^{3}OS), through various doctoral dissertations, workshops, living labs and research projects that have contributed to reorient a number of collaborative practices in the heritage field. Following, these theoretical contributions, two practical examples of participatory governance applied in Ecuador are presented.

The first experience called Orotopía developed by the Technical University of Machala (Utmach), Ecuador. **Iñiguez et al.** relate how the synergy between undergraduate careers affect current systems of participatory governance and public policies on cultural issues in certain municipalities of the province of El Oro, Ecuador. The chapter concludes presenting and analysis and reflection on the influence of the intervention called maintenance campaign on the "Las Herrerías" street in Cuenca – Ecuador according to the four pillars of sustainable development. The more recurrent observations in the participants, were the social and cultural dimension. They affirmed the relationship between neighbors has been improved thanks to this experience (maintenance campaign) and they know each other better and collaborate in participatory activities such as the minga. Furthermore, the inhabitants increase awareness about the tangible and intangible values of their buildings and their neighborhood.

3.2 *Cultural heritage as a source of inspiration for new participatory management approaches*

Within the cultural heritage field, this topic aims to identify: cultural practices, systems of organization,

5

knowledge, among others, which might serve to improve current participatory governance systems. The keynote of this chapter, **M. Lacarrieu**, analyses some experiences that attempted to move towards participatory processes in the heritage field. On that experiences diverse groups from the communities have been included with their resources, logics of action, rules, and systems for shared governance, however, the author the claims those experiences lack reflection around the notions of heritage and culture, commonly built around the idea of the world in terms of inheritance, cultural externality, and separation from social life, with the intention not to subvert its tangibility or its meaning. Therefore, she invites to move towards the idea of inhabiting heritage that involves introducing the relationship between the place and the subjects connected to it, fostering its social appropriation, otherwise a participatory perspective can be a promising road for the field of heritage, but it can also be extremely complex, depending on who can activate the power.

Following this contribution, and taking the case of the WHS of Cuenca, **F. Cardoso & G. García** present some reflections about the sense of cultural heritage. Going in line with Lacarrieu, these authors stress the call to think about committed solutions that include the marginalized (nations, peoples, communities) and the value of the "mestizaje". Concerning to this, the authors underly the need to consider heritage assets adopting an integrated perspective for their conservation, especially including the more modest ones, since each building preserves a history, a history of life, taste, and wisdom expressed in technology and particularly, in the case of Cuenca, the mestizaje -of cultures- which is also evident in the most intimate and sublime expressions of culture and spirituality of the city.

A. Astudillo & E. Acurio bring the analysis of Agenda Ciudadana of Cuenca as an example of participatory management model and an action-oriented model that links cultural and heritage management and civil society participation. According to the authors, on this experience converge the diverse ideologies and needs of the communities that conform the community of Cuenca. However, they stress a lot of work is still needed specially to manage heritage resources in different ways than the ones considered until now.

L. Herrera-Montero, J. Amaya & A. Tenze close this chapter offering a synthetic analysis of capitalism as a global system which engenders negative effects on cultural heritage. Moreover, and based on hermeneutics, the authors provide insights to improve governance processes of cultural heritage, for the protection, preservation and promotion of cultural heritage assets. In this regard, it is referred the approach to cultural heritage as a symbolic legacy for the cultural identity of the peoples, proposed by Prats. This perspective entails a link of its tangible and intangible components, without which any issue of social reality cannot be explained with enough rigor and sense of integrality.

3.3 Lessons from territorial participatory management for effective participatory governance systems in the cultural heritage field

Based on the experiences of participatory management, on the different dimensions of the territory, such as environmental, economic or social, this topic aims to elucidate principles that can be extended to participatory governance in the cultural heritage field. The keynote of this chapter is **J. Farinós** who presents the arguments and proposal to develop smart specialization strategies based on cultural heritage resources. That leads to innovation and supply differentiation (of heritage, landscape, tourism ...) based on high-order advantages. He stresses the opportunity embodied in cultural heritage to correct inadequate socio-economic dynamics with serious negative environmental impacts, allowing territories to obtain the benefits of tourism but minimizing its impact. It means, to configure sustainable heritage destinations, effective cooperation mechanisms between sectors and stakeholders, such as an ethical code for all the actors concerned.

C. Ost & R. Saleh share part of the results of a one-year empirical research during which perception mapping was exploited for analyzing and visualizing "attributed values" based on individual/collective memory in relation to the perceived cultural heritage. Perceptions mapping was carried out in tandem in four partner building/cities/region of the CLIC pro-ject consortium: Rijeka (Croatia), Salerno (Italy), Pakhuis de Zwijger (Amsterdam, the Netherlands), and Vastra Götaland Region (Sweden). The highlight this potentiality of this process to engage and to create por-active attitude among citizens. Indeed, the process of perceptions mapping is a sense-making process during which people map what they feel their cultural, natural and human assets are; express their opinions, ideas, needs and aspirations but also; raise concerns and highlight conflicts related to the management, conservation and preservation of the cultural heritage for future generations.

Going in line with the idea that the conservation of the urban heritage is no longer the preserve of the cultural elite, instead is considered an opportunity to attaining more inclusive, safe, resilient and sustainable cities, **E. Rojas** discusses the opportunities emerging from this trend in Latin America and suggests reforms to integrate the traditional urban heritage preservation and urban planning practices. He proposes three major reforms: (1) transfer the responsibility for the urban heritage from national culture-oriented institutions to local urban development agencies, (2) make urban heritage preservation regulations and interventions an integral part of urban development plans and city management procedures, and (3) incorporate all interested social actors into the urban heritage preservation governance mechanisms by establishing institutional arrangements that promote cooperation and shared responsibility among this diverse set. Being aware that reforms are not easy to implement, he calls to urgently overcome long established bureaucratic traditions and deal with long-established mistrusts and conflicts among social actors in Latin America.

The contribution of **M. Siguencia** presents an interesting analysis -in parallel- between spatial planning and heritage management. Taking as case of study the city of Cuenca, the author evidence on that evolution of processes a kind of convergence of both approaches in the notion of Historic Urban Landscape, particularly on the call to involve different actors. Despite her contribution acknowledge the spaces created by academia, as well as international cooperation activities, to stimulate citizen participation in heritage projects within the historic area, it calls to articulate to public and private it is desirable that, at the CHC, citizen participation re-emerges in a leading role with, for example, a regular citizen observatory present in all sessions, not only narrowed towards historic areas but also consider the city as a whole.

C. Zurbriggen & S. Juri present the experience of Design a transition lab for Resilience and Sustainability Studies that might be considered a source of inspiration for cultural heritage field as an experimental space for transdisciplinary and trans-sector collaboration with multiple actors from academia, public and private sectors, as well as civil society. Transition Design seeks to foster initiatives that can conform ecologies of actions – synergies – to support or disrupt system configurations that can be more appropriate and desirable. Such practical outcomes which may include material and symbolic interventions. The authors identify at least four main challenges link to their proposal: (1) to generate the capacity for anticipation in managing uncertainty; (2) to generate capacity to synthesize knowledge in a transdisciplinary way; (3) to generate the capacity to experiment, to develop tangible spaces in the current context that allow change and; (4) to innovate in the way of evaluating and monitoring processes, adopting a new paradigm oriented to learning, innovation and adaptation in dynamic and complex systems.

Finally, **J. López, D. Pulla & J. Carvallo** present the experience of el Sigig, a rural area in the south region of Ecuador, on which expert-led methodologies were combined to facilitate the effectiveness of citizen participation. It moves a step forward to democratize and legitimate urban decision-making in the Ecuadorian Andes. At the end of the experience, the experts' panel and the population agreed in the decision of the best way to achieve urban revitalization in Sigsig taking into account their cultural heritage legacy.

4   REOCURRING OBSERVATIONS

As described in Section 2, participatory governance in cultural heritage has the potential innovate current government structures and in the cultural heritage field to improve management and protection of cultural heritage. In this regard, the central observation is the call to adopt a systemic view where everything is somehow interconnected. Many authors agree on the significance advances given por example between spatial planning and cultural heritage management that converged in the notion of Historic Urban Landscape. Among other advantages, overcome the division between the material and immaterial nature embodied in cultural heritage assets.

It comes with a second observation, the awareness about a clear conceptual framework. Indeed, some papers claims the need to start participatory processes developing a theoretical reflection. It will serve as the common ground among the diversity of actors and the role each of them plays in the way to reach a *real* participatory process.

In line with these two previous observations, several contributions have shown the multiple opportunities one can find on combining methods and adopting transdisciplinary to create multilevel collaboration among the diverse actors. In the way, capacities are strengthened and integrated in favor of governance policies for cultural heritage and beyond. The common denominator for these contributions is the recognition of the engagement and the participation of the civil society as key aspect to innovate current governance practices. It goes in line with smart specialization policies that emphasizes the importance of the Quintuple Helix Model that includes government, businesses, research/education and civic actors. Some of the papers have remarked the crucial role academic actor might play on participatory governance, acting as a kind of bridge among other actors and their diverse interests.

5   FUTURE RESEARCH

Since the early 2010s, several international and European policies strongly stress the potential of cultural heritage in the sustainable development discourse (CHCFE Consortium 2015), calling for more participation as a key aspect to be reinforced. In this sense, some of the experiences presented in this volume illustrate the attempts to materialize the theory of participation in the cultural heritage field, however, participatory governance is still more a desirable than an existing reality in many places.

Research is still needed to conduct more in-depth assessments to estimate the impact – or not – of participatory experiences derived from the cultural field in the individual interests and short-term perspective that use to characterize each actor. It also includes the capability to managing expectations and the public interest represented by different actors in the governance process, in the Latin American context and beyond.

The prominence of silo-structures characterizes the existing urban and cultural governance in Latin America. Participatory governance of cultural heritage requires capacity strengthening, but also capacity building to construct a new order. It includes the search of organizational measures to deal with the

7

scarcity of financial resources, and updating of current legislation. For future research, it would also be important to examine the most efficient mechanisms to favor interactions among government, academia, industry and civil society considering particularities of each contexts. This would be highly important in order to strengthen heritage-led innovation and thus to contribute to inclusive and sustainable development elsewhere.

REFERENCES

Arnkil, R., Järvensivu, A., Koski, P., & Piirainen, T. (2010). Exploring Quadruple Helix. Tampereen yliopistopaino Oy Juvenes Print: Tampere. *The CLIQ*. Available online: http://urn.fi/urn:isbn:978-951-44-8209-0 (accessed on 5 June 2020).

CHCFE Consortium (2015). Cultural Heritage Counts for Europe, Krakow, Brussels: CHCFE Consortium.

Committee on culture – United Cities and Local Governments – UCLG (2004). Agenda 21 for culture. Bacelona, Spain.

European Union. (2018). Participatory governance of Cultural Heriage. Report of the OMC (Open Method of Coordination) working group of members state's experts.

García, G., Tenze, A., & Achig, C. (2020). The role of the University in maintaining vernacular heritage buildings in the southern region of Ecuador. In Preventive Conservation – From Climate and Damage Monitoring to a Systemic and Integrated Approach.

Garcia, G., Vandesande, A., & Van Balen, K. (2018). "Place attachment and challenges of historic cities: A qualitative empirical study on heritage values in Cuenca, Ecuador", *Journal of Cultural Heritage Management and Sustainable Development*, https://doi.org/10.1108/JCHMSD-08-2017-0054

Landorf, C. (2011). Evaluating social sustainability in historic urban environments. *International Journal of Heritage Studies*, 17(5), 463–477.

McLoughlin, J., Sodogar, B. & Kaminski, J. (2006). Economic valuation methodologies and their application to cultural heritage. In: J McLoughlin, J., Sodagar, B. & Kaminski, J. (eds). *Heritage impact. Proceedings of the first international symposium on the socio-economic impact of cultural heritage*. (pp. 8–27). Budapest: EPOCH.

Paño, P., Rébola, R., & Suárez, E. (2019). Procesos y Metodologías Participativas. Reflexiones y experiencias para la transformación Social. Available online: http://biblioteca.clacso.edu.ar/clacso/gt/2019 0318060039/Procesos_y_metodologias.pdf (accessed on 02 July 2020).

Vandesande A., Van Balen K., Della Torre S., & Cardoso F. (2018). "Preventive and planned conservation as a new management approach for built heritage: from a physical health check to empowering communities and activating (lost) traditions for local sustainable development". *Journal of Cultural Heritage Management and Sustainable Development*, 8(2), 78–81. doi: 10.1108/JCHMSD-05-2018-076.

Vandesande, A., & Van Balen, K. (2019). On established built heritage profiles and capacities, the rise of new profession(al)s and the need for training and education to deal with ongoing disruptive changes in the sector. In: Van Balen, K., Vandesande, A. (Eds.) *Professionalism in the Built Heritage Sector. (Reflections on Cultural Heritage Theories and Practices, 4)*. Boca Raton, FL: CRC Press (Taylor & Francis Group).

Vidovic. (2018). Do it Together. Practices and Tendencies of Participatory Governance in Culture in the Republic of Croatia. Published in the framework of the 'Approaches to Participatory Governance of Cultural Institutions' Project implemented with the support of UNESCO International Fund for Cultural Diversity. Available online: http://participatory-governance-in-culture.net/ (accessed on 10 August 2020).

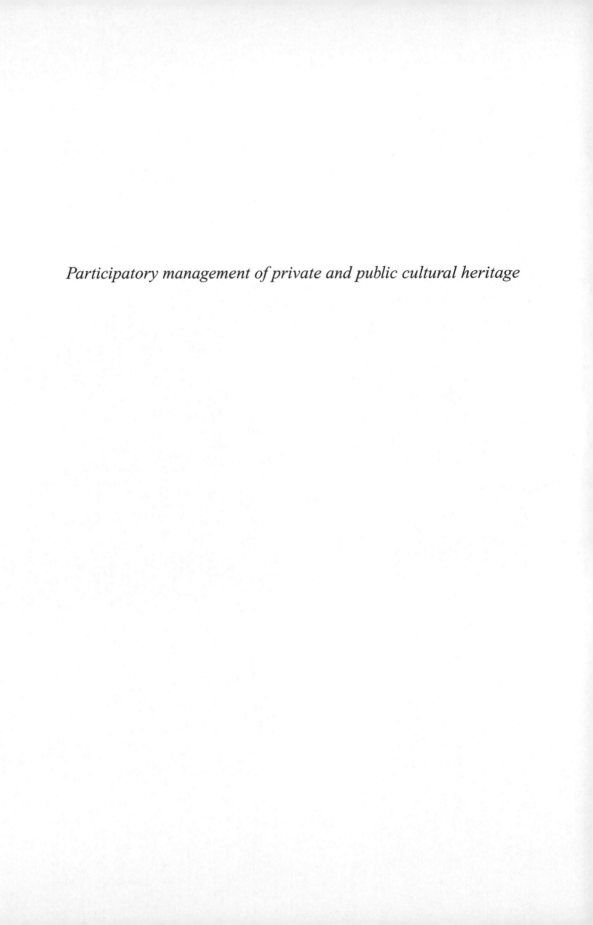

Participatory management of private and public cultural heritage

The Future of the Past:
Paths towards Participatory Governance for Cultural Heritage – García et al (eds)
© 2021 Taylor & Francis Group, London, ISBN 978-1-032-02129-4

Collaborative governance for our heritage: Opportunities and challenges

M. Craps
KU Leuven, Campus Brussels, Belgium

ABSTRACT: Heritage is viewed in this paper as a result-in-the making of an ongoing social process between actors living in different socio-economic conditions with different ways of making sense of their environment. When people identify with heritage, it stimulates them to take care of their environment, of the others and of themselves. This is an opportunity for society. As governments are not able to resolve on their own the complex and value laden questions related with heritage, collaborative governance is necessary. This concepts refers here to the capacity to steer the broad domain of heritage with the active participation of private, community and academic actors. It draws the attention to the importance of dialogue and trusting relationships between actors with diverse legitimate interests. This implies new roles and responsibilities of these actors towards heritage. Shared value creation, community empowerment and transdisciplinary knowledge co-creation are presented as necessary complements for collaborative governance.

1 INTRODUCTION

Let me start with a personal experience that has influenced deeply my own understanding of the importance of collaborative governance for heritage conservation. When I presented a key note inaugural lecture for the "Future of Past" Conference in the old cathedral of Cuenca in December 2019, I realized that exactly 20 years before, on the adjacent central Parque Calderón, I was participating together with my family and thousands of local inhabitants in a big celebration because the historic center of Cuenca was incorporated in the UNESCO list of World Heritage Sites. Indeed, after a long and intensive process in which numerous governmental as well as civil society actors were actively involved, a huge popular 'fiesta" burst out. I took part in it with the same enthusiasm as the local people, because there hasn't been another place in my life with which I identified so intensively as with Cuenca, where I lived for 10 years. I was inspired and adopted by the local people and felt deeply connected with the natural and cultural environment. The local society cultivates permanently its connection with the environment in daily practices and many celebrations, through which it confirms and reinforces a sense of belonging. The strong identification with a shared spatial reality resulting from these practices and celebrations is not monolithic and exclusive for one social group, but it allows the inclusion of different socio-economic and ethnic groups. Even a "gringo" newcomer like me could become part of it!

In what follows I will argue that the active participation of actors belonging to different sectors of society in what I will call "collaborative governance", based on a common and inclusive identification, is of paramount importance for heritage conservation. According to my human and social science perspective, heritage is an important aspect of our lived space, which is the result-in-the making of an ongoing social process: how do we create and re-create this spatial reality through interactions between groups and communities living in different socio-economic conditions and with different ways of conceiving and making sense of the world? Among all these groups we need to construct a common home, our society. This common home is in permanent change and development, between a past, that leaves us a rich material and immaterial heritage, and a future, in which we project our desires and yearnings.

After this introduction I will first explain the threats and challenges for cultural heritage in terms of diversity and resilience, and indicate the importance of heritage as a source of innovation for adaptive and resilient societies. The main part of this paper is focused on the role of collaborative governance to realize the innovative potential of heritage. After clarifying the distinction between government, governance and collaborative governance and the necessity of the latter in a complex world, I point to potential contributions and bottle necks for public and private actors and communities in collaborative governance for heritage conservation. Attention is also given to the specific role of the academic sector in transdisciplinary research for the co-creation of solutions in close collaboration with the other actors. I conclude with underscoring the importance of continuous dialogue and collaboration between public, private, community and academic actors concerning our heritage for a sustainable future.

DOI 10.1201/9781003182016-2

2 HERITAGE, RESILIENCE AND INNOVATION

2.1 *Heritage and resilient societies*

Our cultural heritage, likewise our natural heritage, is suffering serious threats. As for our natural heritage we know that once a species is extinguished, the loss for biodiversity is definitive. There is no way that it will return in existence. This determination is all the more worrying as we are not dealing with isolated species, but different species conform complex eco-systems in which they depend on each other for their survival. When species are extinguished massively – as we are observing in many places of the world currently – this phenomenon diminishes the diversity of the eco-system. As a consequence the system becomes more vulnerable, less resilient to external stress, and due to adverse conditions it may completely collapse.

Something similar happens with our cultural heritage. Each demolition of a meaningful building or historical site is definitive, as we cannot re-create it afterwards. However, we should not consider these building and sites as isolated entities. Their meaningfulness is grounded in being part of socio-cultural and socio-natural assemblies. They are charged with symbolic meanings that help people feel at home and to make sense of their environment (Taçon & Baker 2019). When we demolish without discrimination our heritage from the past, we could find ourselves in an impoverished world, in which it is hard for people to develop a meaningful life. Societies become then more vulnerable.

Empirical studies of the National Trust of England have demonstrated that visitors of heritage sites, for whom these sites have a special meaning in their lives, report an increase in their well-being after their visit (National Trust 2019). Heritage sites just as natural sites give people a feeling of connectedness. In this way they stimulate also people to respect and take care of their environment, the others and themselves, and hence to strengthen the society.

2.2 *Heritage as a source of innovation*

A plea to take care of our heritage is not to be understood as a conservationist invitation to stay nostalgically in a supposedly glorious past. On the contrary, it is meant as an invitation to take our heritage from the past as a source of inspiration for the future. Our heritage is a mirror that allows us to reflect in an appreciative though also critical way on the past, to imagine a desirable future. The name of the conference held in Cuenca, Ecuador, in December 2019, "El futuro del Pasado", suggests that the future needs indeed the past to generate a new future.

However, it's somewhat misleading to speak about "the past", as if there exists just one version of the past. It would be more convenient to speak about "the pasts" in plural. The past is by definition an interpretation, in function of how we are living now and what we aspire for the future. As different human groups experience their lives and environment very differently, and have very different ambitions and visions for the future, they also perceive and interpret the past in very different ways. This happened to our ancestors in former times as well, obviously. They also interpreted and re-interpreted their heritage in function of their present and future, at that time. These interpretation and re-interpretation processes are often clearly reflected in the construction layers of buildings that have been build and re-build in different eras, with different ideas of what is esthetical, useful and valuable.

This multiplicity of criteria to make sense and use of our heritage, does not only occur over time, it occurs as well simultaneously as a result of the unavoidable clash of ideas and interests between different social groups in societies. Moreover local heritage is also indebted to cultural exchange processes, incorporating ideas that may come from different even remote places, as a result of cultural exchange, migrations and even colonization. This may of course cause conflicts between different groups concerning the meaning and use that should be given to certain heritage elements. Groups that historically have been discriminated, may be humiliated by elements that others consider as valuable heritage, and they may demand the de-colonization of this heritage.

3 COLLABORATIVE GOVERNANCE OF HERITAGE

If there are competing values, meanings and interests in play regarding our heritage, a fundamental question is then: how do we decide and act together as a society about what has to be preserved as heritage and in which way? How can heritage be transformed in a bridging resource, stimulating creativity and innovation instead of division? How can different groups and sectors of society be mobilized and involved in this endeavor? These are all fundamental questions when dealing with our heritage. I will argue that collaborative governance, involving actively actors that belong to different sectors of society, is necessary for this endeavor. First I will explain the concept of collaborative governance, as I use it in this paper, by distinguishing it from the concepts of government and governance. Then I will explain more in detail the potential contribution and pitfalls of being involved in collaborative heritage governance for different sectors of society separately.

3.1 *Government, governance and collaborative governance*

The most convenient way to clarify the concept of "governance" is by comparing it with and distinguishing it from the concept of "government", which is much more common in daily language. The government concept puts emphasis on its institutional character that grants legal power to persons and entities from a hierarchical position. From the

'80-ies of the former century onwards this government concept has been criticized for establishing bureaucratic inefficiencies (Osborne 2006). These critiques have originated initiatives with the aim of increasing the "governance" of governments by transforming them in more efficient, transparent and responsible entities (Rhodes 1996).

In practice these governance initiatives have often resulted in weaker governments, with less competences and resources. They were promoted by international financial institutions, worried by the increase of external debts of many governments especially in the global South. Governmental reforms were then used as a leverage for privatization, even in areas of public interest like education, health, transport, infrastructure, etc.

This is clearly not the governance concept I propose in this paper. For that reason I will use the concept of "collaborative governance" to distinguish it from the formerly presented governance concept characterized by reforms "within the government", transferring functions and services to the private sector to enhance its efficiency. Collaborative governance goes "beyond the government". It refers to the capacity to steer broad societal domains, with the active participation of different sectors of society, not only public and private actors, but also including civil and community actors (Huxham 2000; Lownden & Skelcher 1998). Together they co-govern complex issues of broad public interest. In contrast to the concept of government, which has connotations of institutional, legal and hierarchical functioning, the concept of collaborative governance draws the attention to networks of communities with legitimate shared interests, and it is associated with other concepts as participation and co-creation (Crabbé et al. 2018; Fung 2006). Collaborative governance has the aim to go beyond the formal structures and procedures of traditional governments, by setting up adequate social processes and cultivating high quality relationships between the interested actors that favor necessary societal changes (Bouwen & Taillieu 2004).

Collaborative governance refers thus to the capacity of aligning actors and joining resources for issues that are of general interest for society. It is not meant for administrating routine issues but rather to direct change processes at society level when profound changes are deemed necessary for a desirable and sustainable future. (Ansell & Gash 2007; Folke et al. 2005).

The concept of collaborative governance emerged as a reaction to a governability crisis, experienced in different ways in different parts of the world. This governability crisis can be attributed to the hypercomplex, "wicked" nature of the problems with which governments are confronted (Rittel & Webber 1973). Economic-financial and socio-environmental challenges are intertwined with technical and technological questions and with considerations about cultural and ethical values, whatever important topic on the political agendas currently. In these cases governments

can not reunite on their own the legitimacy, resources and varied expertise that would be needed to come up with a solution unilaterally, and they have to appeal to other sectors in society to deal with them jointly.

A fundamental insight of systems thinking is that complex systems can not be steered from the outside, as if they were simple, mechanic systems – like an automobile for instance. Complex systems are emerging and developing "living" entities, that can only be dealt with in an adaptive way from within (Richardson et al. 2004). Nobody - not the government, nor the private entrepreneur, nor the local community or civil society actor and nor even the scientific researcher - can step outside society, which is the system that has to be governed.

3.2 The changing role of the public sector in collaborative governance for heritage

Collaborative governance "beyond the government" does not imply weakening governments, in contrast to the original concept of governance in the '80-ies and '90-ties. On the contrary it requires governments with even more legitimacy than before to enable the accomplishment of new roles and functions, like convening, coordinating, facilitating and supporting other actors with their initiatives to counter societal challenges.

Other actors that do not belong to the public sector have to assume new roles and functions as well, to take part in and contribute to societal governance. New concepts are emerging in different sectors of society to express this shift towards more social responsibility and a bigger societal vocation, like "shared value creation" in the business world (Porter & Kramer 2011); "community empowerment" in civil society (Agrawal & Gibson 1999) and "transdisciplinary knowledge co-creation" in the academic world (Polk 2015). These concepts are complementary and need each other mutually. (Figure 1).

In what follows I will explain that with regards to heritage the transition from governments to collaborative governance requires a private sector that has shared value creation as a vocation, instead of worrying exclusively about financial gains for the owners or stockholders of the firms. Collaborative governance requires also well informed and engaged civil actors

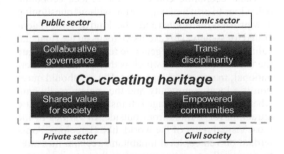

Figure 1. Co-creating heritage between different societal sectors with blurring boundaries.

and social movements, that are critical but also open for collaboration with actors of other sectors. And it requires academic centres and researchers in the service of society, that can co-create knowledge in dialogue between different disciplinary specialists and with different types of knowledge, practices and – often ancient – wisdoms. As a consequence distinctions between different sectors become blurred, and new types of actors appear with "mixed" characteristics, like "social enterprises" or "citizen science" or "community-municipality consortia" (Craps et al. 2004).

3.3   *The contribution of the private sector to the collaborative governance of heritage*

As governments all over the world are confronted with tightened budgets, and the maintenance of heritage demands high investments, they tend to look to the private sector for financial support. As the private sector often disposes of more resources than the public sector, there seem to be indeed attractive possibilities. In this respect we have to consider that part of our heritage, although of public value, in fact is already in private hands and an even larger part could become private property.

Let's take the case of the medieval city of Bruges, a well-known touristic destination in my home country Belgium, to highlight the opportunities and challenges related to the involvement of the private sector in heritage conservation. In this case the municipal government has the intention to sell part of its own heritage assets to generate financial resources for the restoration and maintenance of other buildings and heritage goods, that are considered of more emblematic value for the city. This raises the question if such a privatization is a threat for heritage. With a private sector only thinking and acting in the short term for its own gains, the answer is probably yes. However, with an "alternative" private sector of entrepreneurs that are really interested in the value of heritage, in what it means and symbolizes for the people living daily with it and for society at large, of which the entrepreneurs are part, such a privatization could play an important role in the maintenance of heritage.

This implies that we have to leave behind neoclassical economic concepts, that reduce economic behaviour to profit maximization without questioning how profit is distributed and what kind of value is generated for society. This exclusive focus on the financial value that heritage can generate tends to overestimate the income of tourism, which is then preferably international, massive and high-cost. But we should question if a local community and the society really gain when its complete heritage is transformed in a touristic attraction for visitors from outside. In various of the most visited parts of the world, like Barcelona, Amsterdam or Venice, local inhabitants reclaim measures to counter the negative impact of the massive touristic invasion. One can't minimize these impacts and reduce them to externalities that should be be internalized. How should this impact be compensated when the original population can't continue living in the places where it used to live as a consequence of commercialization and gentrification, and when these place have lost their spirit and meaning?

So we need new economic concepts and models for a private sector in the service of our heritage. Fortunately alternative economic concepts are developed nowadays that are also very promising for the governance of heritage, based on sharing, "peer-to-peer" (Bauwens et al. 2019) and circularity (Ellen McArthur Foundation 2014). For instance in circular economy models objects are not discarded after their consumption, but they contain value from the moment they are produced, during their time of use, which is preferably as long as possible, with possibilities to be repaired and restored, and still afterwards when they are re-used or recycled, without generating waste or negative environmental impact. But…which sector can present better examples to the circular economy than the defenders of our heritage? It is precisely the heritage sector that is interested in preserving and actualizing the value of our material and immaterial production, by giving use and functionality to it as long as possible, by restoring it with craftsmanship and recycling whenever possible, and always with great attention for the social and natural context of heritage sites.

3.4   *The importance of empowered communities for the collaborative governance of heritage*

Collaborative governance cannot be limited to public-private partnerships, but it has to include also the civil society, consisting of local neighbourhood organizations as well as other productive, activist and creative collectives. Empowered communities probably play even the most important role when it comes to successfully implement collaborative governance for heritage (Taçon & Baker 2019).

The cultural heritage of a community is not only its buildings and objects, but also the meanings, ways of living and organizing associated with and implicit in these objects. Heritage preservation can thus not be limited to preservation of stones, walls and buildings, but it must also pay attention to the meanings and values that its inhabitants and users attribute to it. Consequentially important decisions about heritage cannot be taken by public officers, in their offices far away from the living reality, or they cannot be taken based only on private interests. They have to be taken in close collaboration with the people and communities that live in it and use that heritage, through an intimate dialogue about what should be preserved and what should be renovated, and what functions the heritage can fulfil in the future.

In this way heritage has an enormous potential to empower communities through the appropriation of their own living space. This may well generate a virtuous circle: involving communities in the preservation of their heritage is a way to empower them, which will lead to the appropriation of their living space, which will empower them even more. When we aim at empowering communities, their involvement in heritage preservation obviously can't be limited to delivering cheap labour force. They have to be involved in all the stages of a project, from the studies and planification onwards, including capacity building in order to stimulate their appreciation of the artisanal and aesthetic value of the heritage, their identification with it and their motivation to continue maintaining it on its own account.

But what do we mean here with "empowered communities"? The community concept is indeed used in different contexts with a wide variety of meanings (Wenger 1998). On the one hand the community concept refers to local communities, often rural or indigenous communities, that share a determinate geographic context and that identify with a shared history and a vision for the future (Craps et al. 2004). This kind of communities has often a connotation of being "traditional".

On the other hand the community concept refers to informal groups that maintain close and intimate relations among their members, based on a shared interest or activity, like in the case of professional, learning or amateur communities (Wenger 1998). Their members do not necessarily share a geographical space, and their shared history and future vision can be quite limited in time compared to the formerly described local communities. This type of communities, in contrast to local communities, has often connotations of innovation and change.

However the distinction between local communities, as entities of tradition and conservation, and shared interest communities, as entities of innovation, is misleading. In both types of communities informal relations among its members and identification with shared experiences that inspire joint activities, are equally important. Precisely these community characteristics are needed to give life and meaning to heritage sites. Both historically rooted communities and more recently established "creative" communities can contribute in a complementary way to a local dynamic. The first can profit of the newcomers to stimulate their adaptation to broader societal changes, whereas the latter can profit from the first to "localize" their innovative practices.

3.5 The necessity of transdisciplinarity

Heritage experts, with different disciplinary backgrounds, can contribute to the collaborative governance of heritage in various ways. But that implies a transformation of the academic sector as well. Collaborative governance is in need of experts who are aware that they don't have a monopoly on the truth and who are willing to come down from their pedestal to dialogue with actors of the other sectors on equal terms. The expert view "from the outside" on the complex reality of heritage, has to be complemented with the view of the inhabitants and users "from within".

But academic experts will not have to dialogue only with external, non-academic actors, they will also have to dialogue with experts from other disciplines. This is all but evident practice at the universities, in which faculties tend to constitute impregnable and impenetrable fortresses (Craps 2019). Indeed, to capture the complexity of heritage realities, all disciplines that can contribute meaningfully, are needed: architects, historians, archaeologists, economists, urban planners, political scientists, social psychologists, anthropologists, and possibly also others depending on the situations and questions at hand. They have to constitute transdisciplinary learning communities with public, private and community actors about the heritage of common interest.

4   CONCLUSION: DIALOGUING ABOUT THE PAST FOR A SUSTAINABLE FUTURE

The presented model of collaborative governance for heritage may look quite utopic. It is certainly not evident to put in practice this model in many heritage contexts characterized by social conflicts that have resulted in a legacy of prejudices and grudges between the different involved social groups. Moreover the characteristics of the different sectors and their mutual relations, that were presented as conditions for collaborative governance, are still rare in practice: governments still tend to impose their top-down decisions and to consider participation as a necessary evil; many entrepreneurs still tend to focus exclusively on their own financial gains; community actors often fail to appreciate their own heritage, and academics tend to priorate publication success among their disciplinary peers. Each actor shows a tendency to look for his own interests on the short term, without caring for the consequences of his own behaviour on the other actors at system level, and on the long term.

How can we then start acting in a collaborative way, if there is so much conflict, inequality and increasing polarization in our societies? Certainly the collaborative governance of our heritage is still more a desirable than an existing reality in many places, but it is not a utopia neither. It is rather a reality-to-be-constructed by all together, because it is a necessity if we really care about the importance of heritage for our societies. An open dialogue between all interested actors is of utmost importance in this process. When it comes to heritage, we – the actors of our current times – have to be well aware that we are only the passing transmitters of the enormously rich, beautiful and valuable heritage that we have inherited from our ancestors and that we have to pass on to the future generations. Conceiving ourselves in this humble way is helpful to engage in collaborative processes to co-create a sustainable future out of our heritage from the past.

REFERENCES

Agrawal, A. & Gibson, C. 1999. Enchantment and Disenchantment: The Role of Community in Natural Resource Conservation. *World Development,* 27(4), pp. 629–649.

Ansell, C. & Gash, A. 2007. Collaborative governance in theory and practice. *Journal of Public Administration Research and Theory,* Volume 18, pp. 543–571.

Bauwens, M., Kostakis, V. & Pazaitos, A. 2019. *Peer to peer. The commons manifesto.* s.l.:University of Westminster Press.

Bouwen, R. & Taillieu, T. 2004. Multi-party collaboration as social learning for interdependence: developing relational knowing for sustainable natural resources management. *Journal of Community and Applied Social Psychology,* Volume 14, pp. 137–153.

Crabbé, A., Bergmans, A. & Craps, M. 2018. Participation in spatial planning for sustainable cities: the importance of a learning-by-doing approach. In: U. Azeiteiro, et al. eds. *Lifelong Learning and Education in Healthy and Sustainable Cities.* Geneva: Springer, pp. 69–85.

Craps, M. 2019. Transdisciplinarity and sustainable development. In: W. Leal, ed. *Encyclopedia of sustainability in higher education.* Geneva: Springer International Publishing.

Craps, M., Dewulf, A. & Santos, E. 2004. Constructing Common Ground and Re-creating Differences Between Professional and Indigenous Communities in the Andes. *Journal of Community and Applied Social Psychology,* Volume 14, pp. 378–393.

Ellen McArthur Foundation. 2014. *Towards the circular economy: accelerating the scale-up across global supply chains.* Geneva: World Economic Forum.

Folke, C., Hahn, T., Olsson, P. & Norberg, J. 2005. Adaptive Governance of Social-Ecological Systems. *Annual Review of Environment and Resources,* 30(1), pp. 441–473.

Fung, A. 2006. Varieties of participation in complex governance. *Public Administration Review,* Issue December, pp. 66–75.

Huxham, C. 2000. The challenge of collaborative governance. *Public Management,* 2(3), pp. 337–357.

Lownden, V. & Skelcher, C. 1998. The dynamics of multi-organizational partnerships: an analysis of changing modes of governance. *Public Administration,* Volume 76, pp. 313–333.

National Trust. 2019. *Why places matter to people,* s.l.: s.n.

Osborne, S. 2006. The new public governance. *Public Management Review,* 8(3), pp. 377–387.

Polk, M. 2015. Transdisciplinary co-production: Designing and testing a transdisciplinary research framework for societal problem solving. *Futures,* pp. 110–122.

Porter, M. & Kramer, M. 2011. Creating Shared Value. How to reinvent capitalism and unleash a wave of innovation and growth. *Harvard Business Review,* Volume 87, pp. 62–77.

Rhodes, R. 1996. The new governance: governing without government. *Political Studies,* Volume XLIV, pp. 652–667.

Richardson, K., Goldstein, J., Allen, P. & Snowden, D. 2004. *Emergence, complexity and organization.* s.l.:ISCE Publishing.

Rittel, H. & Webber, M., 1973. Dilemmas in a General Theory of Planning. *Policy Sciences,* 4(2), pp. 155–169.

Taçon, P. & Baker, S. 2019. New and emerging challenges to heritage and well-being: a critical review. *Heritage,* 2(2), pp. 1300–1315.

Wenger, E. 1998. *Communities of practice. Learning, meaning and identity.* Cambridge: Cambridge University Press.

The Future of the Past:
Paths towards Participatory Governance for Cultural Heritage – García et al (eds)
© 2021 Taylor & Francis Group, London, ISBN 978-1-032-02129-4

The Xth anniversary of the PRECOM^{3}OS UNESCO chair

K. Van Balen
Director Raymond Lemaire International Centre for Conservation, KU Leuven, Heverlee, Belgium
Holder UNESCO chair on Preventive Conservation, Monitoring and Maintenance of Monuments and Sites

ABSTRACT: Launched in 2009, the UNESCO chair on Preventive Conservation, Monitoring and Maintenance of Monuments and Sites has contributed to the reflections and practices on preventive conservation of the built Heritage. This contribution summarizes the insights gained on preventive conservation mainly through the various doctoral dissertations that have been prepared within this collaboration between the Universidad de Cuenca (Ecuador), the Universidad de Oriente (Cuba) and KU Leuven. A number of workshops, living labs and research projects have contributed to reorient a number of practices to facilitate the implementation of preventive conservation –as is the case with the proposal to include base-line data and monitoring aspects at the introduction of proposals for World Heritage nomination. Experimenting maintenance practices in rural and urban areas in Ecuador has demonstrated they can be implemented. The preventive conservation concept developed has also led to understanding of the contribution of cultural heritage to local sustainable development, and doing so, in creating a new paradigm on the role of cultural heritage for development as has been reported in the report "Cultural Heritage Counts for Europe". The insights seems also to align with a series of other discourses, demonstrating that the role of cultural heritage in society today cannot be underestimated. Prevention helps to understanding the fundamental role of cultural heritage for well-being as it empowers communities to be part of the safeguarding and development of their and next generation's living spaces.

1 INTRODUCTION

The UNESCO Chair on Preventive Conservation, Monitoring and Maintenance of Monuments and Sites (PRECOM^{3}OS) started in 2008 and was inaugurated in Leuven in March 2009 followed by a second inauguration in Cuenca (Ecuador) in December 2009. The chair was a continuation of activities and synergies that had started with European SPRECOMAH project. VLIR-UOS (www.vliruos.be) supported the partnership between KU Leuven (Raymond Lemaire International Centre for Conservation) and the Universidad de Cuenca in Ecuador and the Universidad de Oriente (Cuba) in a ten years long intensive collaboration. Other projects and the Fund of Janssen Pharmaceutica helped to implement various research activities together with the early mentioned universities, Monumentenwacht Vlaanderen and other partners in many different countries.

The collaboration has allowed to improve insights in the fundamental aspects of preventive conservation.

Seven doctoral dissertations and a series of workshops contributed to the creation of incremental insight on the potential of preventive conservation and on the role of cultural heritage in sustainable development.

These insights allows us to sketch the actual context in which we understand the fundamental role of cultural heritage for well-being as it empowers communities to be part of the safeguarding and development of their and next generation's living spaces.

2 PREVENTIVE CONSERVATION ESSENTIALS

2.1 *Preventive Conservation: A definition*

For the sake of proper understanding of the next chapters, we take over essential concepts which we have developed more at length in a previous article on the findings of preventive conservation five years ago (Van Balen 2015). We argued that preventive conservation in the field of built heritage has a distinct meaning and represents a different practice than in archaeology –where it is rather linked to rescue archaeology. It also represents a different practice compared to vast experience with preventive conservation in museums. The latter deals with the conservation of objects for which an optimum (micro) climate can be created to assure optimum preservation conditions. The definition of preventive conservation issued by ICOM-CC (New Delhi 2008) and used in the field of movable heritage reads as: "*Preventive conservation (are)- all measures and actions aimed at avoiding and minimizing future deterioration or loss. They are carried out within the context or on the surroundings of an item, but more often a group of items, whatever their age and condition. These measures and actions are indirect – they do not interfere with the materials and structures of the items. They do not modify their appearance."*

These approaches differ from dealing with historic buildings and sites as (preventive) conservation in this case does usually not allow to change or optimize the environmental conditions, as climatic conditions or exposure to earthquakes, in which the object have to be preserved. Neither is it possible to limit measures to those that "…do not interfere with the materials and structures of the items. …do not modify their appearance …" It is also clear that the circumstances for (preventive) conservation are not limited to responding to urgent actions that threaten historic buildings or sites to disappear. Above that the overall aim of preservation of built heritage is to preserve as much as possible the different heritage values within its social context.

The attention for prevention and maintenance in the field of conservation of monuments and sites is not re-cent although it is gaining attention. The Charter of Athens (1931), article 4 of the Venice Charter (1964), the recommendations from the Council of Europe on Maintenance, preventive actions and crafts (1981) but also the Burra charter refer to the role of maintenance for heritage and society (Cebron 2008). A few common denominators can be found in the arguments given in these documents: preventive conservation helps con-serving authenticity as it avoids or minimize the in-crease of damage thanks to early maintenance and – if necessary- some interventions. From this it is usually deduced that preventive conservation is cost effective. The use of buildings and the proper integration in society enhance the chances for good maintenance. Experiences with monumentenwacht in The Netherlands and in Flanders support the arguments that a preventive conservation approach empowers society at large to take care of its heritage by maintaining it. It is also found that it widens the responsibility for preservation to a larger fraction of society than traditional conservation practices do.

2.2 *What preventive medicine or public health can help us understand about preventive conservation*

In the field of heritage preservation the "medical analogy" is often used to explain methods and approaches. Most practitioners agree that prevention is better than cure, an expression that is used in daily life related to health as well.

Yet in 1971 in a publication on preventive medicine dr. P. Mercenier questioned whether "Preventive Medicine" or "Public Health" is a remedy for all diseases or a requirement in health policy. The article aimed at understanding the limits of curative medicine and investigated the way how the sequential hurdles towards preventive medicine application could be overruled. It builds on the general accepted statement that "Prevention is better than healing". It questioned

whether preventive medicine should replace curative medicine and whether the one is in opposition with the other. It also argues that preventive medicine is about promotion of health (Van Balen 2015).

Preventive medicine identifies different levels of prevention; which over time have been identified slightly differently when comparing the previously mentioned article with that of R. S. Gordon in 1983 (Gordon 1983). The first author identifies two levels of prevention while the second identifies three levels. Primary prevention refers to means to avoid the causes of the un-wanted effect (degradation of health). Secondary prevention refers to means of monitoring that allow an early detection of the symptoms caused by unwanted effects. Finally tertiary prevention refers to means that allow avoiding further spread of the unwanted effect or the generation of new unwanted (side) effects. Referring to the heritage field the latter refers to the "compatibility criteria" (Teutonico et al. 1996) which defines compatibility as avoiding damage to what should be preserved.

How can those findings from the public health sector be "translated" to the heritage field?

The first striking finding is that in the medical world when using the words "public health" and "curative care" these words cover the meaning of the "good state" versus the action to "recover". The meaning also covers the "public nature" of taking care of health. In the conservation field we have no word that expresses that "good state". We could use the world "health" in a wider sense and talk about "heritage health", as is done in the engineering field, e.g. in the field of "structural health monitoring".

Some of the concepts used in public health have yet found their way to the field of built heritage management as in Prof. Stefano della Torre (della Torre 2010) article in which he identified that preventive conservation should be based on three levels of prevention using analogies with medicine. In line with Gordon (Gordon 1983) he defined three levels of prevention as follows:

1. **Primary prevention** that aims at avoiding the causes of the unwanted effect (loss of heritage values or damage) to act;
2. **Secondary prevention** relates to monitoring aiming at early detection of the symptoms of the unwanted effects (loss of heritage values);
3. **Tertiary prevention** relates to avoidance the further spread of the unwanted effect or the (re)generation of new ones.

He clarified that in conservation of built heritage primary intervention starts with assuring the proper use of the building, besides other types of preventive measures as assuring good air quality and good state of maintenance. It should also include a good integration of the heritage in society as to avoid vandalism or neglect and it should be rooted in the regional context.

The importance of the efficacy of monitoring (secondary prevention) has been addressed various times within the meeting organized by the

PRECOM³OS UNESCO chair and its network. Organizations as Monumentenwacht in The Netherlands (www.monumentenwacht.nl) and in Flanders (www.monumentenwacht.be) help owners and site managers to monitor the state of preservation of their heritage assets through regular and reliably updated reports (Stulens 2002; Verpoest et al. 2006). More than 25 years of Monumentenwacht experience in The Netherlands and in Flanders have been able to demonstrate that "prevention is better than cure". Referring to the comments give on the article of dr. P. Mercenier, in the field of public health, monitoring should not only be considered a treatment but mere a diagnostic tool. The application of such monitoring tools, as we learned from the medical analogy, should consider a balance between time and resources for analysis and for monitoring versus the speed of decay progress. It should understand the risks and traps of screening systems (e.g. false negatives, false positives) when conclusion are taken (Van Balen 2015).

As in preventive medicine or public health, preventive conservation requires a system consisting of a logic set of organizations and policies that work towards the goal of preserving the good state of conservation of the heritage building stock. It is different from a policy that aims at responding to repairing (avoidable) damage to individual buildings. There are relevant organisms that contribute to such a systemic approach as Monumentenwacht, "Civil protection" or "fire brigade". In Italy for example, the Civil protection had a major contribution in understanding the possible risks and impacts of earthquakes on heritage and they also developed emergency measures in case of earthquake events. Civil protection can be considered part of a preventive conservation system as it exists independently of the "individual" need for repair of a monument, at the condition that it has expertise of dealing with the specificities of heritage. While in some countries as in Italy this organization has developed this expertise, in other countries such organizations may lack the required sensibility for built heritage.

A state of the art of the understanding and implementation of preventive conservation has been shared in previous contributions (Van Balen et al. 2013; Van Balen 2015).

In analogy with studies on preventive medicine three conditions for the implementation of a prevention approach can be given. Firstly scientific knowledge should be available to understand "causes", to understand the socio- economic context. To be able to implement such programs and their efficacy operational research is really needed. Secondly there should be properly trained professionals and finally the required preventive services (the system) should be available.

Related to the first point, the activities of the UNESCO chair as well as diverse initiatives developed by its members have helped to boost the creation of scientific knowledge in the field as we will report in the next chapter. However still a lot is to be done. Referring back to the preventive conservation of movable objects, for the reasons mentioned before we argue that its practice is still very much object based and that a wider, encompassing systemic is lacking. There might be a great benefit to align the preventive conservation concepts for movable and immovable heritage.

The experience of the involvement of Civil Protection in Italy (Modena 2010) shows how this organization can be involved in emergency care but at the same time can contribute to preventive care by coordinating damage assessment and learning from past earthquake events.

The analogy with medicine instructs us that most of the professionals in the field of curative heritage conservation are trained to see "the monument or isolated site" as the demander of the care. In preventive conservation or (public) "heritage-health care" it is the system that should provide all necessary service to the individual building or to heritage building stock. As, to be effective, preventive conservation should address a large part of the heritage stock or at least of the group of more vulnerable heritage buildings. Professionals should learn thinking in term of the "heritage building stock" (society) and not in term of separate buildings or sites (individuals) and therefor professionals should be trained to think in terms of risks and to address them in a systemic, integrated way. It reminds us many things we are confronted with in today's COVID19 crisis.

As a consequence, we argue that a "(built) Heritage's health care system" should be designed to support in an effective way preventive conservation. Such a system should assure a good health for the heritage "stock" as a whole and should be less concerned with resolving individual needs only.

It will not be easy to introduce such a system and to make it acceptable due to the delayed perception of the advantageous effects of preventive measures. As today people and decision makers tend to choose for the fastest result, the delayed results of preventive measures should be compensated by the superiority of its outcomes above the results of curative actions. The objective elements that help making the choice for preventive conservation strategy are often difficult to define by lack of convincing examples, which explains partially the lack of interest for prevention. Even Monumentenwacht organisations with a nice proof record have difficulties in getting the necessary support from decision makers. Monitoring and follow-up maintenance actions by the owners based on the report of Monumentenwacht during those years has for example resulted in an improvement of the state of preservation of the gutters and roofs of the stock of monitored heritage buildings in Flanders. Better alignment between monitoring and maintenance interventions is needed something which we have tried to address in one of the doctoral research projects explained in the next chapter. Indeed while thanks to the existence of Monumentenwacht in Flanders a partial preventive conservation "system" exists that supplies monitoring capacity, within a legal framework and (unfortunately eroded) incentives, the "system" is incomplete. One of the bottlenecks that has been identified in a survey carried out by Monumentenwacht Vlaanderen (Monumentenwacht 2010) is the lack of available companies that can carry out maintenance work or small restoration work based on their report. The lack of trained craftsmen and the lack of integration of related activities in the local economy contribute to that bottleneck.

Therefore, although the prevention approach seems advanced in Flanders, the systemic approach is not sufficiently developed or understood to have a fully working preventive conservation system available for immovable heritage in Flanders. Similarly we argue that in a country as Italy which has a civil protection organization that is rather well developed, the existence of such an organization is not sufficient for a preventive conservation system.

Preventive conservation is complementary to curative conservation and both should be integrated in a system of promotion and support of "heritage health". Similarly as in medicine preventive conservation should always prevail, keeping in mind that curative action –in a way- is always a defeat. In analogy with public health we can conclude that the benefits of preventive conservation are not only the arguments used since more than ten years that it that it improves the preservation of heritage values and its cost-effectiveness but that additionally its potential to involve a larger part of the population in the process, should be acknowledged. Indeed in the case of preventive conservation the responsibility for follow-up and for repair actions is much more local which involves more people.

3 INSIGHTS GAINED OVER TEN YEARS

During the last ten years various PhD dissertations have been made that along with the related publications and the publication of various workshops make up the body of knowledge that has been created through the collaboration within the PRECOM³OS UNESCO chair.

3.1 Doctoral dissertations resulting from the PRECOM³OS UNESCO chair

While the research has been the result of inter-university collaboration all PhDs have been made and registered at KU Leuven involving the chair holder. The last cited PhD has led to a joint PhD with the Universidad de Oriente in Santiago de Cuba.

Most of the doctoral research support has been given through the VLIR-UOS which supports partnerships between universities and university colleges, in Flanders and the South, that are searching for answers to global and local challenges (www.vliruos.be). The institutional cooperation with the Universidad de Cuenca, that included collaboration on World Heritage Cities Management lasted 10 years starting in 2008. The institutional collaboration with Universidad de Oriente in Santiago de Cuba started in 2013 and will last until 2022. It includes a Project that focuses on social and cultural local development and heritage preservation in the Eastern Region of Cuba. A VLIR-UOS South-South-North project "Latin-American axe on preventive conservation of built heritage" in 2015–2016 facilitated the mutual exchange between the three involved universities, including direct bilateral cooperation between the projects in Ecuador and Cuba. Other doctoral research was sponsored by the Janssen Pharmaceutica Fund on Preventive Conservation that has supported the chair during 10 years from 2007 on. The other funding sources for the doctoral research and the related projects came from the Belgian Science Policy Office, the Flemish research fund and from a collaboration with the Flemish Immovable Heritage Agency.

3.1.1 Towards a 3D GIS Based Monitoring Tool for Preventive Conservation Management of the World Heritage City of Cuenca

The First PhD terminated within the activities of the chair was the doctoral dissertation by Dr. Veronica Heras, in collaboration with Universidad de Cuenca: "Towards a 3D GIS Based Monitoring Tool for Preventive Conservation Management of the World Heritage City of Cuenca" (Heras 2014). This research focused on the technological aspect of monitoring the state of preservation of built heritage at the level of a World Heritage City using Geographical Information Systems (GIS). This GIS became part of the management instrument of the World Heritage city of the Historic Centre of Santa Ana de los Ríos de Cuenca. The research made a first attempt to develop a 3 dimensional version of GIS, allowing to use 3D functionality in the monitoring of the values and the state of preservation of World Heritage. The publication "A value-based monitoring system to support heritage conservation planning." (Heras et al. 2013) gives a good insight in the outcome of this work.

3.1.2 Integrated Approach in Management of Coastal Cultural Heritage

The doctoral dissertation of dr. Sorna Khakzad, "Integrated Approach in Management of Coastal Cultural Heritage" (Khakzad 2015) may not have been directly connected to the activities of the chair, however it revealed a methodology to connect different types of cultural heritage that are "separated" by different conventions while reality requires them to be managed jointly. In her research, that got the support from a project which involved also the Flemish Cultural Heritage Agency, she studied the interface of different cultural heritage regimes (below and above water) by addressing how to integrate the intermediate coastal cultural heritage in the integrated management of coastal zones. The eco system service approaches seemed very relevant to integrate cultural heritage, nature and societies in such management. and services approach of the ecological system. Also the introduction of the middle ground concept in the negotiation of various interests (different communities, nature versus culture, …) has been inspiring. The article "Coastal cultural heritage: A resource to be included in integrated coastal zone management" (Khakzad et al. 2015) is a good and well-cited publication on her work.

3.1.3 Preventive Conservation Strategy for Built Heritage Aimed at Sustainable Management and Local Development

The doctoral dissertation of dr. Aziliz Vandesande "Preventive Conservation Strategy for Built Heritage Aimed at Sustainable Management and Local Development" (Vandesande 2017) resulted in the outline of a preventive conservation strategy, which entails a relevant and new approach to implement innovations and changes in the existing built heritage sector. This strategy is based on 2 fundamental building blocks, i.e. (1) the Preventive Conservation System (PCS) which structures the impact of a preventive conservation approach on sustainable management and local development and (2) the Multi-Level Perspective (MLP) or quasi-evolutionary model based on sociotechnical innovations and evolutionary economics which demonstrates emergent innovation patterns and dynamics. (Vandesande 2017). The preventive conservation strategy is approached starting from understanding local sustainable development which induces a better understanding of the systemic needs that preventive conservation requires to be implemented and to remain sustainable. The research also describes the societal background changes that clarifies gradual changes in approaching the preservation of cultural heritage since last century. It also identifies the components that may contribute to a regime change and to the implementation of a preventive conservation model. The publication "Preventive and planned conservation as a new management approach for built heritage: from a physical health check to empowering communities and activating (lost) traditions for local sustainable development." (Vandesande et al. 2018) contributes to framing this debate properly.

3.1.4 Integrating Monitoring in the Nomination Process of Cultural Serial Transnational World Heritage Using Geospatial Content Management Systems

The doctoral research of dr. Ona Vileikis on "Integrating Monitoring in the Nomination Process of Cultural Serial Transnational World Heritage Using Geospatial Content Management Systems: The Silk Roads Case Study." (Vileikis 2018), was carried out in the context of a project sponsored by the Belgian Science Policy office in support of the activities of UNESCO. The project supported preparatory work for the nomination of various Silk Road related sites in Central Asian Countries on the World Heritage List. It developed an on-line platform allowing various countries to jointly nominate corridors of the Silk Road. The dissertation defined a conceptual framework for strengthening the monitoring process of cultural serial transnational World Heritage nominations supporting

a preventive conservation approach, while also building stakeholder capacity. It implemented and tested the conceptual framework using the case study of the Silk Roads serial transnational World Heritage Corridors in Central Asia. The contribution of the PhD to improving preventive conservation is highlighted in the proposed inclusion of base-line data that onsets the monitoring of World Heritage Sites starting from the nomination process and this not waiting for the first periodic reporting. It also integrated maintenance aspects into the World Heritage application process, anticipating periodic reporting and including maintenance management. The publication "Monitoring and measuring change: The Silk Roads Cultural Heritage Resource Information System (CHRIS), presented at the 18th ICOMOS General Assembly Scientific Symposium. "Heritage and Landscape as Human Values", gives a good understanding of the outcome of this research (Vileikis ct al. 2014).

3.1.5 The Activation Process of Cultural Heritage as a Driver of Development

The doctoral research carried out by dr. Gabriela Garcia Velez, on "The Activation Process of Cultural Heritage as a Driver of Development" (Garcia 2018) was integral part of the VLIR-UOS project on World Heritage Cities' Management. The doctoral dissertation addressed two main critical questions: "Can culture contribute to mitigating threats to development?" and "If so, how could culture be integrated into development strategies?".

A validated theoretical framework was built to comprehend the dual role of culture as a structural dimension of development and as an agency for mobilizing the other dimensions of sustainable development as defined in the report Cultural Heritage Counts for Europe report: the social, the economic and the environmental dimensions (CHCfE 2015).

This Activation Process model places culture, cultural heritage at the core and fosters its maximum use as a driver of development while balancing with the preservation of cultural heritage. This Activation Process provides a guide to develop diagnosis and self-diagnosis studies, and even more, to sustain a transparent process of negotiation of change among (multi) actors. A synthetic article on part of this doctoral research is available in the article "Place attachment and challenges of historic cities: A qualitative empirical study on heritage values in Cuenca, Ecuador" (Garcia et al. 2018).

3.1.6 Historic Urban Landscape Approach for the Conservation of the Historic Centre of Cuenca

In her PhD dissertation "Historic Urban Landscape Approach for the Conservation of the Historic Centre of Cuenca, Ecuador" dr. Maria Eugenia Siguencia Avila, (Siguencia 2018) aligned preventive conservation with the 2011 Recommendation on Historic Urban Landscapes' (HUL), integrating tangible and intangible heritage and addressing the needs of the local stakeholders. Her research demonstrated how the 2011 Recommendation aimed at developing effective tools for implementing and ensuring the conservation of urban heritage, contributing not only to the local cultural heritage agenda but also to urban development processes. The holistic understanding from the heritage and from the urban planning field, is implicit in the 2011 Recommendation acknowledging "the need for interactions between a physical environment and the community that is located in it. T the doctoral research therefor focused on the territory, on the activities carried out as well as on the stakeholders involved in the implementation of the HUL approach.

An in-depth study of the case of Cuenca in Ecuador on the implementation of the 2011 Recommendation was carried out within a comparative analytical framework of 102 case studies spread world-wide that had developed various types of activities linked to the HUL approach. The analysis proved that the use of the 2011 Recommendation promoted activities carried out by the academic part of the community but had limited enabling impact on public administration. The analysis allowed to position the Latin American case of Cuenca in a global framework. It demonstrated that the 2011 Recommendation succeeded only partially in the creation of the expected interactions amongst the "territory", the "activities" and the "stakeholders" in Cuenca. An essential article on part of this doctoral research is available in the "Historic urban landscape: an approach for sustainable management in Cuenca (Rey-Perez et al. 2017).

3.1.7 *Quality Improvement of Repair Interventions on Built Heritage*

The doctoral research of dr. Nathalie Van Roy on "Quality Improvement of Repair Interventions on Built Heritage" (Van Roy 2018) was supported by a PhD grant from the Flemish Research Fund and by integrating the research in the project "Changes in Cultural Heritage Activities: New Goals and Benefits for Economy and Society" abbreviated Changes Project (http://www.changes-project.eu/). The latter was a Joint Program Initiative Cultural Heritage project of which the expenses from KU Leuven were supported by the Belgian Science Policy Office.

The most important contribution of this work is the development of a specific quality management approach for built heritage that is founded on a solid scientific basis. The basis is provided by the analysis of current practice and the valorization of two quality improvement methods. Both analyses provided scientific evidence of the importance of quality management, as well as the importance of its specificity for built heritage.

The research had a specific focus on the Flemish context and contains mostly micro-analyses of specific repair interventions. It is a first extensive study on this topic in Flanders, it provides a point of departure and a set of interesting research questions for continuing the research on the quality management of built heritage.

The analysis of current practice provides interesting data on how interventions are managed, designed and implemented in practice. The case study analyses furthermore allowed to identify challenges and opportunities, giving a voice to the daily struggles of owners and professionals.

A variety of implemented research methods were needed to tackle the complexity of the analyzed cases and to detect tendencies and concepts that contribute to a better understanding of conservation practice. From the interaction between the various stakeholders the research revealed the need to establish communities of knowledge and of practice including owners, crafts people, contractors to promote direct exchange amongst stakeholders seeking to respond to the necessary (maintenance) interventions that the monitoring activities of Monumentenwacht had revealed. A good summary of part of the research and illustration of the methodology can be read in dr. Nathalie Van Roy's article "A preventive conservation approach for historical timber roof structures." (Van Roy et al. 2018).

3.1.8 *Preventive Conservation of Historic Urban Areas Addressing Cultural Values and Socio-Economic Dynamics. Case Study of the Vista Alegre District in Santiago de Cuba*

The doctoral research by dr. Luis Bello Caballero, "Preventive Conservation of Historic Urban Areas Addressing Cultural Values and Socio-Economic Dynamics. Case Study of the Vista Alegre District in Santiago de Cuba" (Bello 2019) considered the interrelationship between cultural values and socioeconomic dynamics as essential in heritage preservation. In particular, diverse social and economic phenomena have influenced the decay of historic urban areas in Cuba, leading to a permanent claim for integrated new approaches that can contribute to their preservation. The research relied on the preventive conservation of built heritage as an emerging alternative for sustainable conservation actions since it focuses on the causes of decay to reduce adverse effects on authenticity and integrity. The overall objective was to determine the influence of the inter-relationship between cultural values and socio-economic dynamics on the preservation of historic urban areas, by developing an analytical sequence based on the preventive conservation approach, applied to the case study of the Vista Alegre District in Santiago de Cuba. Understand the principles of preventive conservation assisted in developing a multidisciplinary toolbox approach that structures the analytical sequence to be applied on the case study.

This research's main contribution to knowledge was the integration of the otherwise separated concepts of cultural values and socio-economic dynamics in the context of preventive conservation, and to determine their influence on the preservation of historic urban areas. The results can be considered as a framework that assists the conservation and management of historic urban areas by facing the challenges posed by diverse social and economic phenomena.

Dr. Luis Bello's article "Documenting the Impact of Socioeconomic Dynamics on Heritage Sites. The Case of Vista Alegre District in Santiago de Cuba." (Bello et al. 2017) reports on the documentation aspect of the research carried out in the framework of this PhD.

3.2 Other key-publications

Since the launch of the PRECOM³OS UNESCO chair different events have been organized that resulted in separate publications. Some of them are listed below in chronologic order:

– Cardoso, F., (Ed.) (2012) II Encuentro Precomos Seminario - Taller de Tecnologías y restauración de Obras en Tierra. Cuenca (Ecuador), Proyecto VlirCPM, Universidad de Cuenca
– Paolini, A., Vafadari, A., Cesaro, G., Santana Quintero, M., Van Balen, K., Vileikis, O., Fakhoury, L. (2012). Risk Management at Heritage Sites: a case study of the Petra World Heritage Site. Paris: UNESCO and KU Leuven (http:// openarchive.icomos.org/1456/1/217107m. pdf)
– Cardoso, F. and Rodas Vasquez, C., (Eds.) (2013) Encuentro PRECOMOS - Desafios de la Conservación Preventiva. Cuenca (Ecuador), Proyecto VlirCPM, Universidad de Cuenca
– Van Balen, K. and Vandesande, A., (Eds.) (2013). Reflections on Preventive Conservation, Maintenance and Monitoring of Monuments and Sites by the PRECOM³OS UNESCO Chair. Leuven: ACCO. ISBN: 9789033493423
– (2015). Conferencia Visionaria. Una mirada ciudadana de Cuenca hacia el futuro ¡Todos tenemos algo que decir! Universidad de Cuenca, Ecuador. Available at: http:// whc.unesco.org/document/ 137365.
– Van Balen, K., Vandesande, A. (Eds.) (2015). Heritage counts. (Reflections on Cultural Heritage Theories and Practices. A series by the Raymond Lemaire International Centre for Conservation, 2). Antwerp: Garant Publishers. ISBN: 978-90-441-3330-1. Open Access.
– Van Balen, K., Vandesande, A. (Eds.) (2015). Community involvement in heritage. (Reflections on Cultural Heritage Theories and Practices. A series by the Raymond Lemaire International Centre for Conservation, 1). Antwerp: Garant Publishers. ISBN: 9789044132632. Open Access
– Della Torre, S. and Borgarino, M., (Eds.) (2016) Proceedings of the International Conference Preventive and Planned Conservation: Vol. 4. International Conference Preventive and Planned Conservation. Monza-Mantua, 5–9 May 2014., Milan: Politecnico di Milano e Nardini Editore.
– Van Balen, K., Verstrynge, E. (Eds.) (2016). Structural Analysis of Historical Constructions - Anamnesis, Diagnosis, Therapy, Controls. Leiden, The Netherlands: CRC/Balkema. ISBN: 97811380 29514.

– Achig, M. C., Garcia Velez, G. E., Cardoso, F., Vázquez, L., Jara, D., Barsallo, G., and Rodas, T. (2017). Campaña de mantenimiento de las edificaciones patrimoniales de San Roque 2013–2014. Universidad de Cuenca
– Cardoso Martínez F. (Ed.). (2017). Discursos y experiencias para la gestión del patrimonio Febrero-Marzo 2016. Universidad de Cuenca. ISBN: 978-9978-14-367-4.
– Rey Pérez J., Astudillo Cordero S., Siguencia Avila ME., Forero J., Auquilla Zambrano S. (Eds.) (2017). Historic Urban Landscape The application of the Recommendation on the Historic Urban Landscape in Cuenca - Ecuador. A new approach to cultural and natural heritage. Cuenca - Ecuador: Universidad de Cuenca. ISBN: 9789978143513.
– Van Balen, K., Vandesande, A. (Eds.) (2018). Innovative Built Heritage Models. (Reflections on Cultural Heritage Theories and Practices, 3). CRC Press/Balkema (Taylor & Francis group). ISBN: 9781138498617.
– Van Balen, K., Vandesande, A. (Eds.) (2019). Professionalism in the Built Heritage Sector. (Reflections on Cultural Heritage Theories and Practices, 4). CRC Press (Taylor & Francis group). ISBN: 0367027631.

4 CHALLENGES TODAY

From an actual perspective and considering the past experiences and outcomes of the activities of the PRECOM³OS UNESCO Chair we invite the academic community but also the communities of practice in the field of cultural heritage, to reflect on a number of challenges in the field of heritage preservation that may need specific attention today and in the near future.

The first one relates to understanding the mechanism that drives our gradual changes in concepts and approaches in dealing with cultural heritage.

The second one relates to understanding the essential role and contribution of cultural heritage in society.

The third one relates to positioning preventive conservation of built heritage and sites in this actual cultural heritage scenery.

4.1 Understanding what drives changing concepts on preservation of cultural heritage

Different theories and models explain the regular changes of concepts that influence the way we are reviewing and innovating our understanding of the world and activities which are occurring. Evolution of our understanding and the way humanity act, are usually not seen as linear "progress" but influenced by waves and period of chaos in which new insights can surface leading to paradigm shifts as explained yet by Thomas Kuhn in his famous publication "the structure of scientific revolutions" in 1962. In the field of economics Kondratieff and Schumpeter "wave" models are most know and often referred to. Chaotic periods or periods of transition to new regime systems (Loorbach 2007) as the one we are living in today are opportunities to notice new drivers and new equilibrium that were hidden by the normative existing regimes. Today consequences about how we have been dealing with nature, with acceptance and driving towards inequality are surfacing in the response again COVID19. It shows that -yet for some time- changes in perspective are occurring, also in the field of cultural heritage.

The "Culture 3.0" concept popping up in the beginning of the twenty-first century elucidates the transformation of seeing participating people not (only) as audiences but as practitioners and actors. It is the consequence of changes in socio-technical regimes that acknowledge the social and economic value creation through culture (Sacco, 2018). The upstream approach in the Cultural Heritage Counts for Europe report (CHCfE, 2015) can be read in a similar way and evidences the contribution of cultural heritage to sustainable development. We also remind here that the implementation of preventive conservation benefits from public participation.

4.2 The role of (built) cultural heritage in societies today

"The report "Cultural Heritage counts for Europe" (CHCfE, 2015) made by the consortium with the same name under the leadership of Europa Nostra, in which RLICC has been privileged to contribute to, has helped me to understand to what extend the preservation of cultural heritage is about connecting society with its cultural heritage. Not only –does it seem- is society at large not all sufficiently aware of the value of that heritage to our society, the challenge is also how that (undervalued) cultural heritage can contribute to the aspirations of society" (Van Balen, 2018) Also in various of the above mentioned research outcomes, it becomes clear how influential and needed Cultural Heritage is for people's wellbeing. The report "Places that make us" (National Trust, 2017) and the more recent report "Why places matter to people" (National Trust, 2019) show how cultural heritage contributes to experiencing elements that make up the "Five Ways to Wellbeing": connecting to others, being active, taking notice, keep learning, and giving to others and society. It is striking how much of those findings echo with the significance given to cultural heritage in the Special Eurobarometer Cultural Heritage (European Commission, 2017).

Those reports show that being involved in cultural heritage is about "giving to others and society" (National Trust, 2019) and that "Large majorities think cultural heritage is important to them personally, as well as to their community, region, country and the EU as a whole" (European Commission, 2017), demonstrating that cultural heritage is about shared values (including their diversity!) and that it is relevant to seek for ways the common interest of people can benefit from innovative governance models

as –for example- advocated by the P2P foundation (https://p2pfoundation.net/).

4.3 Position of preventive conservation

Many of the above findings can be integrated in developing improved preventive conservation strategies and models. Essential in these strategies and models remains the prevailing of prevention above curing; the integration of cultural heritage in the discourse of sustainable development keeping cultural heritage values at the core, while understanding the role of cooperation of stakeholders for their well-being benefits. The broader the involvement of society the easier the societal benefit can be elucidated stimulating broad participation, which eventually (as is the case in public health) will drive efforts more in the direction of maintenance and prevention than in (avoidable) cure. Prevention helps to understand the fundamental role of cultural heritage for well-being as it empowers communities to be part of the safeguarding and development of their and next generation's living spaces.

5 CONCLUSIONS

The applicability of preventive conservation requires a system wide approach. In analogy with "public health" strategies and applications, components of such a preventive system have been sketched. Seven doctoral dissertations and a number of workshops have allowed to further develop those insights and to evaluate a number of practical applications. They have helped to understand the contribution of cultural heritage to sustainable development and to position (preventive) conservation in safeguarding cultural heritage and in improving society's wellbeing.

We recognize how interest in this approach coincides with broader societal evolutions which impact society's perception of the role and contribution of cultural heritage. Prevention helps to understanding the fundamental role of cultural heritage for well-being as it empowers communities to be part of the safeguarding and development of their and next generation's living spaces. New governance systems should be developed for that purpose.

REFERENCES

Bello Caballero, L., (2019). Preventive Conservation of Historic Urban Areas Addressing Cultural Values and Socio-Economic Dynamics. Case Study of the Vista Alegre District in Santiago de Cuba. Leuven: KU Leuven. Engineering Science, Open Access.

Bello Caballero, L.E., Van Balen, K. (2017). Documenting the Impact of Socioeconomic Dynamics on Heritage Sites. The Case of Vista Alegre District in Santiago de Cuba. In: ISPRS Annals of the Photogrammetry, Remote Sensing and Spatial Information Sciences, (31–38). Presented at the 26th International CIPA Symposium 2017,

Ottawa, Canada. doi: 10.5194/isprs-annals-IV-2-W2-31-2017 Open Access

Cebron Lipovec, N., Van Balen, K. (2008). Practices of monitoring and maintenance of architectural heritage in Europe: examples of "Monumentenwacht" type of initiatives and their organisational contexts. In : Kolar, J. (Eds.). CHRESP: Cultural Heritage Research Meets Practice, Lubljana, 10-12/11/2008: pp. 238–239.

CHCfE Consortium (2015). Cultural Heritage Counts for Europe: - funded by the EU Culture Programme (2007–2013). (available at: blogs.encatc.org/culturalheritage countsforeurope/outcomes/).

Della Torre S. (2010). Critical reflection document on the draft European Standard CEN/TC 346 WI 346013 Conservation of cultural property- Condition survey of immovable heritage, unpublished discussion document, Seminar on condition reporting systems for the built cultural heritage, Monumentenwacht Vlaanderen, 22-24/02/2010.

European Commission (2017), Special Eurobarometer Report 466: Cultural Heritage, https://epale.ec.europa.eu/en/resource-centre/content/special-eurobarometer-report-466-cultural-heritage

Garcia Velez, G. (2018). The Activation Process of Cultural Heritage as a Driver of Development. Leuven: KU Leuven. Engineering Science, Open access.

Garcia, G., Vandesande, A., van Balen, K. (2018). Place attachment and challenges of historic cities: A qualitative empirical study on heritage values in Cuenca, Ecuador. Journal of Cultural Heritage Management and Sustainable Development, 8 (3), 387–399. doi: 10.1108/JCHMSD-08-2017-0054

Gordon R.S. (1983). An operational classification of disease prevention, Jnl. Public Health Rep. Mar-Apr 1983; 98(2): 107–109.

Heras, V. (2014). Towards a 3D GIS Based Monitoring Tool for Preventive Conservation Management of the World Heritage City of Cuenca, Leuven: KU Leuven. Engineering Science.

Heras Barros, V.C., Wijffels, A., Cardoso, F., Vandesande, A., Santana Quintero, M., Van Orshoven, J., Steenberghen, T., Van Balen, K. (2013). A value-based monitoring system to support heritage conservation planning. Journal of Cultural Heritage Management and Sustainable Development, 3 (2), 130–147. doi: 10.1108/JCHMSD-10-2012-0051

Khakzad, S. (2015). Integrated Approach in Management of Coastal Cultural Heritage. Leuven: KU Leuven. Engineering Science, Open Access.

Khakzad, S., Pieters, M., Van Balen, K. (2015b). Coastal cultural heritage: A resource to be included in integrated coastal zone management. Ocean & Coastal Management, 118, 110–128. doi: 10.1016/j.ocecoaman.2015.07.

Loorbach, D. (2007). Transition management: new mode of governance for sustainable development, PhD dissertation, Rotterdam (Erasmus Universiteit Rotterdam).

Modena C., Casarin F., da Porto F. and Munari M. (2010). L'Aquila 6th April 2009 Earthquake: Emergency and Post-emergency Activities on Cultural Heritage Buildings in M. Garevski, A. Ansal (eds.), Earthquake Engineering in Europe, Geotechnical, Geological, and Earthquake Engineering 17, Springer, pp. 495–521

Monumentenwacht Vlaanderen (2010). Tevredenheidsenquête 20 jaar Monumentenwacht (Satisfaction Survey (of members) 20 years of Monumentenwacht); www.monumentenwacht.be

National Trust (2017). Places that make us, National Trust, UK, https://nt.global.ssl.fastly.net/documents/places-that-make-us-research-report.pdf

National Trust (2019). Why places matter to people, National Trust, UK, https://nt.global.ssl.fastly.net/documents/places-matter-research-report.pdf

Rey-Perez J., Siguencia Avila ME. (2017). Historic urban landscape: an approach for sustainable management in Cuenca (Ecuador). Journal of Cultural Heritage Management and Sustainable Development, 7 (3), 308–327.

Sacco, P.L.; Ferilli, G.; Tavano Blessi, G. (2018). From Culture 1.0 to Culture 3.0: Three Socio-Technical Regimes of Social and Economic Value Creation through Culture, and Their Impact on European Cohesion Policies. Sustainability, 10, 3923

Siguencia Avila, M., (2018). Historic Urban Landscape Approach for the Conservation of the Historic Centre of Cuenca, Ecuador. Leuven: KU Leuven. Engineering Science, Open Access.

Stulens, A. (2002). Monument Watch in Flanders : an outline, in: A. Stulens, A. (ed.) First International Monumentenwacht Conference 2000, Amsterdam, p. 15.

Teutonico, J.M., et al., Group Report: How Can We Ensure the Responsible and Effective Use of Treatments (Cleaning, Consolidation, Protection)?, Dahlem Workshop on Saving Our Architectural Heritage: The Conservation of Historic Stone Structures, eds. N.S. Baer & R. Snethlage, John Wiley & Sons: Berlin, pp. 293–315, 1996

Van Balen, K. (2015). Preventive Conservation of Historic Buildings. International Journal for Restoration of Buildings and Monuments, 21 (2), 99–104.

Van Balen, K. (2018). Opening address at start of Thematic Week "Professionalism in the built Heritage Sector", February 5th, 2018.

Van Balen, K., Vandesande, A. (Eds) (2013). Reflections on Preventive Conservation, Maintenance and Monitoring of Monuments and Sites by the PRECOM^{3}OS UNESCO Chair. Leuven/Den Haag: ACCO.

Van Roy, N., (2018). Quality Improvement of Repair Interventions on Built Heritage. Leuven: KU Leuven. Engineering Science, Open Access

Van Roy, N., Verstrynge, E., Van Balen, K. (2018). A preventive conservation approach for historical timber roof structures. Journal of Cultural Heritage Management and Sustainable Development, 8 (2), 82–94.

Vandesande, A. (2017). Preventive Conservation Strategy for Built Heritage Aimed at Sustainable Management and Local Development. Leuven: KU Leuven. Engineering Science, Print.

Vandesande, A., Van Balen, K., Della Torre, S., Cardoso, F. (2018). Preventive and planned conservation as a new management approach for built heritage: from a physical health check to empowering communities and activating (lost) traditions for local sustainable development. Journal of Cultural Heritage Management and Sustainable Development, 8 (2), 78-81. doi: 10.1108/JCHMSD-05-2018-076

Verpoest, L., Stulens, A., (2006). "Monumentenwacht. A monitoring and maintenance system for the cultural (built) heritage in the flemish region (Belgium)", in Conservation in changing societies, Heritage and development / Conservation et sociétés en transformation, Patrimoine et développement. Patricio, T., Van Balen, K. and De Jonge, K., (Eds.)., Raymond Lemarie International Centre for Conservation, Leuven, pp. 191–198.

Vileikis, O., Van Balen, K., Santana Quintero, M. (2014). Monitoring and measuring change: The Silk Roads Cultural Heritage Resource Information System (CHRIS). Presented at the 18th ICOMOS General Assembly Scientific Symposium. "Heritage and Landscape as Human Values", Florence.

Vileikis, O. (2018). Integrating Monitoring in the Nomination Process of Cultural Serial Transnational World Heritage Using Geospatial Content Management Systems: The Silk Roads Case Study. Leuven: KU Leuven. Engineering Science, Open Access.

The Future of the Past:
Paths towards Participatory Governance for Cultural Heritage – García et al (eds)
© 2021 Taylor & Francis Group, London, ISBN 978-1-032-02129-4

Participatory governance of cultural heritage at Ecuador: Orotopía

G. Iñiguez, L. Quezada & F. Tusa
Technical University of Machala, Machala, Ecuador

ABSTRACT: This paper systematizes a Project Community Connect called Orotopía at the Technical University of Machala (Utmach), Ecuador, which was constituted to resignify the importance of the cultural heritage in the local territory, through of an interdisciplinary work with several undergraduate careers of the School of Social Sciences at the same university (Utmach). In this article, the authors analyze the contribution of the project in front of the participatory governance in the territory and, at the same time, they analyze how the topic of cultural heritage can be a source of inspiration for new approaches for participatory management. The methodology applied is the systematization of an action research experience in the local territory. Moreover, this analysis evaluates the Orotopía's project organization and how the synergy between undergraduate careers affect current systems of participatory governance and public policies on cultural issues in certain municipalities of the province of El Oro, Ecuador.

1 INTRODUCTION

1.1 *Cultural heritage: An overview*

The cultural heritage consists of manifestations of human life which represent a particular view of life, and witness the history and validity of that view. The expression of culture or evidence of a way of life, may be embodied in material things such as monuments or sites. Archaeological sites and human built-structures are clearly accepted as important evidence of the past to be preserved. Thus, the remains of ancient cities, historical complexes and urban ensembles show the importance of modern life evolution or a now abandoned way of living (Prott & O'Keefe 1992).

The complex relationships between tourism and heritage are revealed in the tensions between tradition and modernity. The role of heritage in postmodern tourism is examined, particularly built heritage, which is at the heart of cultural tourism (Nuryanti 1996). A natural link exists between tourism and cultural heritage management, and yet a debate occurs between them on the sustainability of heritage tourism. What it is also missing, it is a process whereby elements of both areas can be included in the identification and actualization of the tourism potential of cultural heritage place (du Cros 2001).

Cultural and heritage tourism represent a major area of growth in the special interest tourism market. However, despite the tourist popularity of visiting heritage sites and participating in cultural activities, a relatively minimal detailed attention has been given to the cultural and heritage tourism phenomenon.

The travel industry is increasingly recognizing the significance of cultural and heritage resources and their marketability. However, the maximization of the long-term benefits of cultural and heritage tourism requires developing effective management strategies that will ensure the conservation and the proper use of irreplaceable cultural and heritage resources (Zeppel & Hall 1991).

Heritage is a concept to which most people would assign a positive value. The preservation of material culture, objects of art and of daily use, architecture, landscape form, and intangible culture, dance performances, music, theater, and ritual, as well as language and human memory, are generally regarded as a shared common good by which everyone benefits. Both personal and community identities are formed through such tangible objects and intangible cultural performances, and a formation of a strong identity would seem to be a fundamentally good thing. But heritage is also intertwined with identity and territory, where individuals and communities are often in competition or outright conflict (Silverman & Fairchild Ruggles 2007).

Heritage usually comprises those things in the natural and cultural environment around us that we have inherited from previous generations, or sometimes we have created by the current generation, and that we think, as communities and societies, are so important we want to pass them on to the upcoming generations. As previously noted, these things can be tangible (places, artifacts) and intangible (practices and skills embodied in people). Heritage is the result of a selection process. The aim of heritage protection is to pass on this selection of things with their intact values and in authentic condition. Or at least this is how we think about tangible heritage (Logan 2007).

This paper aims to provide a conceptual analysis of cultural heritage ecosystem and how they are linked to

DOI 10.1201/9781003182016-4

the concepts of landscape, identity and local tourism. It discusses how these cultural ecosystem services can be assessed and integrated into spatial and physical planning. Also, this research presents one case study, called Orotopía project, to shed light on the assessment process. A case study from El Oro province, located in Ecuador country, combines an analysis of cultural heritage ecosystem with methods for documenting cultural heritage values in landscapes.

It is important to mention that the main authors of this research have participated as managers of the Orotopía program, through the planning, evaluation and verification of each of its operational phases, since 2015. In this regard, we worked with students and teachers in this project of linking with the community, being the Communication career the promoter of this initiative focused on the line of research on heritage and narrative. The objective was to create communication products focused on making visible the tangible and intangible cultural heritage of the cantons of Chilla, Pasaje, Santa Rosa, Machala and Las Lajas, in the province of El Oro, Ecuador.

This paper examines the nature of the relationship between tourism and cultural heritage management in the established urban destination of El Oro province (McKercher et al. 2005). By mean of the Orotopía research project, we demonstrate that the methods from cultural heritage conservation provide tools for the analysis of historical values as well as historical drivers of change in landscapes that can add time-depth to more spatially focused cultural heritage ecosystem (Tengberg et al. 2012).

2 THEORETICAL FRAMEWORK

2.1 Heritage

Cultures summarize diverse collective experiences of the society. The meanings derived from it, allow the generation of symbolisms and connotations historically inherited by common people (Segura-Lazcano 2017).

Besides, culture refers to the set of creations that emanate from a community, based on tradition, is transmitted by a group or a human collective; they are part of the cultural and social identity, hold on norms and values transmitted orally, by imitation or in other ways. Cultural representations include, among others: language, literature, music, dance, games, mythology, rituals, customs, crafts, architecture and other arts (González-Varas 2015).

In relation to other cultural products, Maldonado (2017: 101) indicates that "heritage is result of a dual, individual and social culture. Those are multiple in nature (material and immaterial) and discipline (historical, artistic, archaeological, oral, documentary). Some heritage that, based on the insertion of ICTs in a non-context, are deterritorialized so that each person can make them their own and serve as a connector between different realities, as a motivating element to explore new cultures".

Likewise, what it is now considered patrimony in the past was an element without any kind of importance. Although, these have been celebrated or existed in the past, they were not taken as patrimony, due to the lack of significant value or acts of appreciation. Currently, these are considered heritage due to a context that gives them a specific historical-cultural value. It is necessary the existence of an institution that promotes an integrated management system of cultural heritage and that takes into account all its dimensions to manage them properly, in an orderly manner, according to the multiplicity of values and socio-discursive practices, in which, they are inserted (López, Encabo & Jerez 2011: 7–8).

The definition of heritage is not only framed in those monuments that represent victorious memories, but also heritage includes the lamentable affairs, losses, the anonymous protagonists, resistances and everything that is part of the integral construction of historical heritage assets. The will to truth and the will to hegemonic visibility can hide and condemn other patrimonies to omission, as has happened in the past (Criado-Boado & Barreiro 2013).

In relation to the heritage that is promoted within the urban areas, it is found that this is not a preexisting material, but rather a social construction in which traditionally the groups that maintain power pick out some buildings and certain historical parts of the city. Thus, assessment refers to the relationships that people and their elites have with their history and their present (Espoz & del Campo 2018: 7–8).

Inside the characteristics of postmodern society, it is found the confluence of several cultures within the same geographic space. This generates as a result the interaction of ethnic groups from internal or external migratory flows. This cultural diversity is understood as a plurality of linguistic minorities, of social classes, of gender and of economic level. In the same way, current conflicts of multiple political, religious, economic and cultural spheres do more complex the dynamic relationship between these multicultures (Gómez, Gil & López 2017: 484).

In front of the dialogue that is required between people who agglutinates various groups of a society appear guidelines for an interrelation between them. In addition, the processes of globalization accentuated by the digital society make more evident the need for an intercultural relationship, which must be located in the local-national-global articulation (Mardones 2015: 12).

2.2 Citizen participation

Speaking of citizenship, it is something more than appealing to an administrative category in relation to a legal recognition that a specific state grants to people. It is to appeal to a condition: that of being autonomous, free from responsibility and protagonist in relation to different spheres or dimensions of public life. It means overcoming the forms of vassalage or submission that can occur in different areas of our life in society (Gozálvez & Contreras-Pulido 2014: 130).

Regarding reflexive citizenship, Thiebaut (1998: 169) indicates that "the citizen is an active reflective subject, capable of operating with a diversity of logics in different systems of action: his conception of the public must be marked by a deep tolerance; it must adopt a reflective attitude with regard to what believes as a condition for assuming his responsibilities". It is a citizen who is in constant construction in accordance with the social and cultural changes of the territories. In this context, citizenship is transformed and consists of different perspectives and narratives (Álvarez-Moreno & Vásquez-Carvajal 2015: 487).

In this regard, it is proposed formation and acquisition of a kind of moral autonomy that generates critical personalities capable of working towards social development and with determined competencies to contribute to the promotion of a citizenship with skills and attitudes that favor social equality (Aguaded-Gómez & Caldeiro-Pedreira 2017: 10).

In recent decades, the social concern for the preservation of cultural heritage has run parallel to a growing interest in nature conservation. In reality, both processes can be considered characteristic of a modernity that is increasingly aware of the vulnerability of cultural and natural resources to the acceleration of history and the constant transformations of technological society (González-Varas 2015).

2.3 Governance and development

Territory is a relevant factor in the way people organize their lives and conceive their identity. It also allows planning and public policy actions to be planned and raises the discussion on the most appropriate territorial scale for these actions (Delamaza & Thayer 2016).

For Olivera (2011), the territory is composed of physical, historical, economic, political and social events, but also of non-visible realities, worthy of being analyzed by geography, since they are capable of marking a space with material traces that reflect their existence and certain human behaviors that have a great symbolic and identitarian meaning that help to form the sense of place.

Starting from the fact that governance refers to the process of interaction and negotiation of interests between heterogeneous actors that determine the form and modalities of taking decisions and exercising power (Stoll Kleemann et al. 2006: 4), the result should be the widespread acceptance and subsequent implementation of decisions taken together in the case of participatory governance or also known as good governance (Brenner & de la Vega 2014).

Governance is a concept of synthesis and synergy as it is defined by Aguilar (2010), which implies the development of a process of directing the society, characterized fundamentally by no longer having a single actor (government), but rather a collective process in which multiple actors participate (Miranda 2016: 153).

It is necessary to recognize and to study the processes of production of economic value related to the theme of heritage. For instance, it is necessary to analyze: what is it the value? how is it established? for the benefit of whom? which management model is profitable? what business plan is acceptable? how do we subtract the patrimonial goods from the process of reification and commercialization? where do we situate the boundary between this dimension of the patrimony and the unlimited tendency to speculation? (López, Encabo, Jerez 2011: 15).

According to the new agenda of cultural policies for development, the cultural heritage "constitutes an invaluable bookmark of the experiences and human aspirations that nourish our daily lives. Therefore, it adds value to the human being and it is for this reason that it deserves to be protected, enhanced and transmitted for future generations". In this sense, the design of such policies should focus on safeguarding it (Zink & Cornelis 2016: 375).

As a result, the state has to become a key agent to achieve the democratization of culture, understood in terms of access and production possibilities (Zink & Cornelis 2016: 376).

2.4 The impacts of tourism on society and culture

In the next graphic (see Table 1) we can observe the positive and negative impacts about the influence of tourism on society and culture, based on the research of Gonzalez-Varas (2015).

Table 1.   Effects of tourism on society and culture.

Positive impacts	Cultural approaches and exchanges, such as an experience that is part of the framework of openness to other cultures and ways of life that contribute to increasing mutual understanding between peoples and respect for ethnic and cultural differences.
Negatives impacts	Depersonalization of tourist/host relationships. Acculturation and globalization.

Source: Gonzalez-Varas (2015).

3   METHODOLOGICAL APPROACH

As a Project Community Connect, Orotopía applies an action-research approach in the local territory at El Oro province, Ecuador, which gave a new meaning to our local understanding about cultural heritage. First of all, the Orotopía project was approved by the Department of society liaisons, cooperation, internships and practices (VINCOPP) after an open call for projects in 2015. Subsequently, it was accepted by the university council at the Technical University of Machala. For the preparation of the text report, Orotopía project applied the technique of bibliographic review and several interviews with the officials of the National Institute of Cultural Heritage (INPC) in Ecuador.

At the methodological level, qualitative-quantitative instruments were applied in order to collect information related to each of the objectives planned within the project. In this sense, we use surveys, heritage registration forms, in-depth interviews, focus groups and working groups with members of the project, the beneficiary community and institutions related to cultural heritage and professionals in this study area.

For the quantitative survey, an instrument was used to obtain information from the intervened sector which was validated by expert judgement formed by the technicians of the Cultural Heritage Institute and Utmach professors who intervene in the program and the Culture Directions of the municipalities. The final objective was explored the following topics:

- Cultural heritage and social inequality.
- Uses of heritage.
- Purpose of preservation.
- Heritage at the time of the cultural industry.
- Affirmation of patrimonial registry.
- Activities to promote cultural heritage.
- Usefulness of cultural heritage.

In order to carry out the surveys, the population of the cantons was taken as a reference established at 363124 inhabitants (INEC 2010). In this regard, the sample was 1064 respondents with an error margin of 3%. About the territory to be intervened, the political division of the cantons: Machala, Santa Rosa, Pasaje, Las Lajas and Chilla and their inhabitants were taken into account.

4 RESULTS

4.1 *Orotopía initial aims*

Under the need of generating development from the local heritage management, the Orotopía project was proposed for contributing and for strengthening the identity and heritage of the areas involved with the project while achieving greater visibility of the various attractions of the sector and generating opportunities for socio-cultural-economic development (Quezada et al. 2017: 1116).

In order to achieve the aspirations of the project, professionals from the following undergraduate majors were called to work together: Communication, Sociology, Plastic Arts, Hospitality & Tourism, Civil Engineering, Systems Engineering, Early Childhood and Nursery Education. In addition, the project included heritage issues in the agenda of the municipalities (Table 2).

a. *Cultural heritage baseline:* The Orotopía project proposes the following axes of promotion through which the local culture and its history are promoted:

- *Faces:* each of these historical processes are immersed some personalities, who by

Table 2. Orotopía initial aims.

Cultural heritage baseline	The diagnosis of the sector registers different patrimonial assets obtaining basic information to be disseminated among population, and thus, to ensure the knowledge transmission for upcoming generations.
Archaeological heritage tourist route	It promotes the economic matrix of the sectors intervened through the population training and diffusion about tourist routes and their attractions.
Visibility in media and multimedia digital spaces	It allows to the population increase its knowledge of the existing goods within the region and promote the productive matrix.

Source: Quezada et al. (2017).

their intellectual qualities of management and leadership, helped to change the course of their territory. In this context, each locality raised the name of one or more leaders who laid the foundations for educational, social, economic and cultural progress. Orotopía, in its commitment to incorporate local cultural heritage, rescued the value of those intellectual references of great courage and effort, contributed to build a fairer and sovereign society.

- *Magic:* the orense cultural heritage is not limited to places and monuments, but goes beyond and addresses aspects of the imaginaries of their people. Preserving this type of heritage is important due to the ephemeral nature. In this axis of promotion, we included: oral traditions, rituals, festive acts, knowledge, ancestral techniques, fantastic, legends, myths and stories of El Oro province.
- *Enchantment:* this axis opens way through the gastronomy of the local towns, where the great variety of typical plates, of the high and low part, are a potential tourist resource in growth and it considers all those monuments and constructions whose historical, aesthetic and anthropological value give account of the transcendence of El Oro province.

b. *Orense cultural heritage registry cards:* The baseline information survey is worked jointly with the INPC Zone 07 (National Institute of Cultural Heritage), this government institution had an old register of heritage assets located in the different sectors intervened by the Orotopía project. The total was 105 records in rural parishes and 117 records in urban areas. This information allowed a mapping of cultural assets. Then, by using the initial data, a declaration of cultural heritage was developed from the areas of Tourism, Civil Engineering, Sociology.

The INPC, under code IBI-07-09-50-000-000001 (August 3 2018), declared as a heritage property to the Coello Theater, which is located in the urban center of Pasaje city. This register is part of the Information System of Ecuadorian Cultural Heritage (SIPCE). The application was submitted in the framework of the Orotopía project by the Municipality of Pasaje, the House of Culture at El Oro province and the Technical University of Machala.

c. *Municipal ordinance:* The Orotopía project applied surveys in the urban area of the cantons of Machala, Pasaje, Santa Rosa, Las Lajas and Chilla. These statistics established the relevance of the incorporation of cultural heritage in the management of territorial development. In addition, the population criteria about heritage assets were identified. Thus, strategies were formulated for the recovery, conservation and promotion of cultural goods with a view to the space construction for collective interaction within these territories.

This study generated strategies for the effective management of cultural heritage, which in turn gave the municipalities of Pasaje and Santa Rosa a substantial basis for the formulation of the heritage ordinance, which was non-existent within these levels of local governments and its development was of utmost importance, since the Territorial Ordinance Law (COOTAD) in Ecuador established, as an obligation, the heritage assets management to each municipality. This law enabled to the national government disburse economic items to local governments for the cultural management benefit.

d. *Tourist route of the archaeological heritage:* For the generation of economic activities related to the heritage use from the Hospitality and Tourism career, four tourist routes are planned and designed based on the archaeological heritage of El Oro province. This project seeks to use, in a sustainable way, the resources identified in the information gathering thanks to the patrimonial registry files. The tourist routes were presented to the INPC for technical validation to allow its operation. In addition, the tourist routes allowed local government tourism offices to have specialized guides, which include signage indicators for easy identification of tourist resources on maps of the province.

e. *Visibilization in media and multimedia digital spaces:* Here, you will find all the communicational products generated by the following careers: Social Communication, Plastic Arts and Systems Engineering. In the first instance, concepts such as: Faces, Magic and Enchantment are applied in the registration files in order to obtain communicational resources as: photographs, oral memory of facts, written memory or documents, videos, music, etc. For the proper diffusion of contents, a diffusion strategy about communicational products based on transmedia narrative is proposed as detailed next in Table 3.

f. *Children's educational contents:* From the research of the students of the Early Childhood and Nursery Education career, a children's literature book

Table 3. Media communicative products.

Radio	60	Radio microcapsules: Legends, character interviews, historical reviews, heritage terminology and interculturality.
	16	2 seasons of radio programs on cultural themes.
Newspaper	8	Series of publications "Cultural landscapes of the rural parishes of El Oro province".
	5	Reports on intercultural themes.
Video	3	Documentaries of Santa Rosa, The Magic of the train, Quinteto de Oro.
	4	Gastronomic reports.
	4	Gastronomic heritage capsules.
	2	Intercultural "Pukyall" competition for secondary and infant education.
Photography	1	Photographic exhibition of "Cultural landscapes of the rural parishes of El Oro province".
	1	Photographic bank.
Social network	1	Facebook (Orotopía), Twitter (@orotopía_), Instagram (@Orotopía), Flickr Orotopía: Crossmedia narrative publications of the 17 cultural landscapes of the rural parishes of El Oro province". Publications of illustrious characters-faces. Ancestral Medicine Publications. Photo galleries. Geolocation of heritage sites.
Itinerant Museum	22	Busts of emblematic characters of the intervened zone within the province of El Oro. Engravings of craftsmanship. Murals of the faces of the community.
Website	1	Web platform with resources for download, consultation, research.
Children's literature book	100	Survey of the baseline of sound and poetic heritage and transference in early education centers.

Source: Orotopía Project (2019).

called "The magic world of Orotopía" that includes a text production of stories, poems, songs, fables and puppet show. The information comes from the local oral memory and an adaptation work of literary narratives. The books were given to the teachers of the educational centers of the intervened zone with the purpose of preparing a second part of the project, which will contain an educommunicative plan about patrimonial literacy.

g. *Prizes and recognitions:* The work for the insertion of cultural heritage into the local government agenda allowed to the Orotopía project to obtain the National Recognition of Good Practices by the National Institute of Cultural Heritage. The project also was declared among the five finalist projects in the UNOS Contest (for society liaisons projects at the national territory). In addition, the canton of Pasaje extended recognition to the Orotopía project because its inter-institutional support with the Cultural Heritage office in Pasaje city, during its CXXIV anniversary of cantonization.

5 CONCLUSIONS

For the development of a project of intervention in territory, it is necessary to organize an interdisciplinary team from diverse areas of knowledge that allow to find solutions to the starting problem which was established thus:

In this sense, seven undergraduate university majors are integrated: Social Communication, Plastic Arts, Sociology & Political Sciences, Initial & Nursery Education, Hospitality & Tourism, Civil Engineering and Systems Engineering. The contribution of the academy to society allowed students to strengthen their professional skills within the generation of multiple functional products, which were supervised by university professors and the INPC national institute.

The management of cultural heritage is only possible due to the existence of a baseline for the heritage assets registration. Otherwise, the current situation of the assets and their possible applications at the tourism, research, economic and cultural level is unknown. In order to achieve this study, it is necessary to construct file formats for registration, to develop the instructions for filling in information, to train those responsible for collecting data; this allows the preservation state evaluation and the goods conservation.

The intervention of the Orotopía project in governance strengthens the organizational and management capacity of public entities in terms of the development of cultural activities with the implementation of strategies for recovery, care and preservation of cultural heritage, both in urban and rural areas, allowing citizens a proper characterization of the potential of heritage wealth, valuation of identity and sense of belonging in the cantons of Chilla, Pasaje, Santa Rosa, Machala and Las Lajas.

One of the results of the project is the incorporation of cultural heritage issues in transversal axes of the educational curriculum in territory. For example, in the early education career, the baseline survey was synthesized, and a book of sound and poetic heritage was created with the collaborative participation of early education teachers and the local community. On the other hand, the students of the Plastic Arts career have intervened in the schools developing representations on iconic characters of El Oro province, construction that starts from the development of an educommunicative planning with the global concept: 'Faces, Magic and Enchantment'.

REFERENCES

Aguaded-Gómez, I., & Caldeiro-Pedreira. 2017. Autonomy or media subordination? Citizen formation in a contemporary communicative context. *Diálogos de la Comunicación* 93.

Álvarez-Moreno, M.A., & Vásquez-Carvajal, S.C. 2015. Radio y cultura: una propuesta de radio ciudadana en Internet. *Palabra Clave* 18(2): 475–498.

Brenner, L. & de la Vega, A. 2014. La gobernanza participativa de áreas naturales protegidas. El caso de la Reserva de la Biosfera El Vizcaíno. *Región y Sociedad* XXVI, 59.

Criado-Boado, F., &, Barreiro, D. 2013. El patrimonio era otra cosa. *Estudios Atacameños* 45: 5–18.

Delamaza, G., & Thayer, L. 2016. Percepciones políticas y prácticas de participación como instrumento para la gobernanza de los territorios. Un análisis comparado de escalas territoriales en la macrorregión sur de Chile. *Eure* 42(127): 137–158.

Du Cros, H. 2001. A new model to assist in planning for sustaina cultural heritage tourism. *International Journal of Tourism Research*.

Espoz Dalmasso, M., & del Campo, M. 2018. Estrategias de comunicación política: sentidos del patrimonio y el turismo en Córdoba (2010–2018). *Question 1(60)*: 103.

Gómez Barreto, I., Gil Madrona, P., & Martínez López, M. 2017. Valoración de la competencia intercultural en la formación inicial de los maestros de educación infantil. *Interciencia 42*(8): 484–493.

González –Varas, I. 2015. *Patrimonio cultural.* Madrid: Básicos Arte Cátedra.

Gozálvez, V,. & Contreras-Pulido, P. 2014. Empoderar a la ciudadanía mediática desde la educomunicación. *Comunicar XXI*(42): 129–136.

Quezada, L., Iñiguez, G., Benítez, K., & Tusa, F. 2017. Difusión del patrimonio cultural, tradiciones y saberes ancestrales de la provincia de El Oro: Orotopía. *Revista Conference Proceeding*.

Logan, W. 2007. *Closing Pandora's Box: Human Rights Conundrums in Cultural Heritage Protection.* Springer Link.

López, A., Encabo, E., & Jerez, I. 2011. Competencia digital y literacidad: nuevos formatos narrativos en el videojuego Dragon Age: Orígenes. *Comunicar XVIII* (36): 165–171.

Maldonado, S. 2017. Heritage education and social media. From research to action: the Heritage Education project. *Revista Pulso.*

McKercher,B., Ho, P. & du Cros, H. 2005. Relationship between tourism and cultural heritage management: evidence from Hong Kong. *Tourism Management* 26(4): 539–548.

Miranda, C. 2016. Gobernar en el Siglo XXI: La Necesaria Consolidación de la Gobernanza. *Sapienza Organizacional* 3(6): 151–166.

Nuryanti, W. 1996. Heritage and postmodern tourism. *Annals of Tourism Research* 23(2): 249–260.

Olivera, A. 2011. Patrimonio inmaterial, recurso turístico y espíritu de los territorios. *Cuadernos de Turismo* 27: 663–677.

Prott, L. & O'Keefe, P. 1992. 'Cultural Heritage' or 'Cultural Property'? *International Journal of Cultural Property* 1(2): 307–320.

Silverman, H. & Fairchild Ruggles, D. 2007. *Cultural Heritage and Human Rights*. Springer Link.

Tengberg, A., Fredholm, S., Eliasson, I., Knez, I., Saltzman, K., & Wetterberg, O. 2012. Cultural ecosystem services provided by landscapes: Assessment of heritage values and identity. *Ecosystem Services* 2:14–26.

Zeppel, H. & Hall, C. 1991. Selling art and history: cultural heritage and tourism. *Journal of Tourism Studies* 2(1): 29–45.

Zink, M.E. & Cornelis, S. 2016. Investigadores y patrimonio: una fructífera experiencia. *Revista Conexão UEPG 12*(3): 374–389.

33

The Future of the Past:
Paths towards Participatory Governance for Cultural Heritage – García (eds)
© 2021 Taylor & Francis Group, London, ISBN 978-1-032-02129-4

Impacts from "Las Herrerias" maintenance campaign on the community

O. Zalamea
KU Leuven/SADL, Leuven, Belgium

G. Barsallo & M.C. Achig-Balarezo
City Preservation Management Project, Faculty of Architecture, University of Cuenca, Cuenca, Ecuador

ABSTRACT: The article explores how a maintenance campaign on the "Las Herrerías" street in Cuenca – Ecuador impacted the community with regard to their perception of the cultural, social, environmental and economic value of heritage. This research aims to analyze and reflect on the influence of the intervention on buildings on the quality of life of the inhabitants and identify the perception of the community about the heritage values of the neighborhood. This article presents the thematic analysis of 18 interviews with owners, tenants, custodians and neighbors of the buildings involved in the maintenance campaign. The opinions of the participants have been organized according to the four pillars of sustainable development. These results have made it possible to generate an analysis of the performance on the site, with the aim of preserving the heritage values of the buildings.

1 INTRODUCTION

1.1 *Preventive conservation of heritage*

The importance of heritage conservation has been evident for some years now. This term from the perspective of ICOM-CC is considered as actions that allow the safeguarding of cultural heritage so that it transcends from generation to generation. Furthermore, conservation is considered to comprise preventive and curative conservation, and restoration (ICOM-CC 2008).

In this context, preventive conservation aims to anticipate the deterioration of heritage assets caused by natural factors or human actions. This is achieved through the application of a set of measures and actions established in the identification of the values of buildings (Barsallo 2019).

It should be noted that the recognition of people, institutions or communities is what defines what should be preserved, based on their reflections and memories that lead to the valuation and preservation of their past (Achig et al. 2017). Thus, from a social perspective, preventive conservation proposes sharing the responsibility for the care of heritage assets (Cardoso 2012). The role played by each actor ensures the sustainability of the preventive conservation process.

1.2 *International context*

Internationally there are several organizations and institutions specialized in the maintenance of heritage buildings:

The UNESCO chair for Preventive Conservation, Monitoring and Maintenance of Monuments and Sites (PRECOM³OS). It was established in 2008 and promotes conferences and debates around the concept of preventive conservation and sustainability (Van Balen & Vandesande 2013).

On the other hand, mention should be made of the successful experience of "Monumentenwacht", a non-governmental organization started in Flanders in 1991 and in the Netherlands in 1973. Monumentenwacht supports the owners and managers of historic buildings to prevent deterioration through a systematic, careful monitoring and by carrying out maintenance work (Stulens 2002). Its actions are focused on generating a guide to the state of conservation of the property and suggesting intervention actions in the building if necessary. Subsequently, owners and managers should contact specialists, for example, restoration architects, structural engineers, contractors, etc. (Vandesande 2016).

Another initiative that is concerned with the conservation of built heritage operates in the Netherlands with the company Stadsherstel. It has intervened in around a hundred buildings and monuments, such as: churches and industrial monuments like pumping stations and a shipyard. Although the purpose of the company is focused on intervening, the maintenance processes are considered essential activities since they ensure the future and the proper functioning of the property so that it transcends its equity value over time. Stadsherstel promotes a maintenance mechanism with shared responsibility between the company and the tenant. These maintenance activities and insurance

DOI 10.1201/9781003182016-5

costs are contractually divided between these two actors in such a way that: large maintenance works are assumed by the company, small ones by the tenant, while the structure is insured by the company and its interior by the tenant (Vermeulen 2012).

1.3 Local context

In Ecuador, at the University of Cuenca, the City Preservation Management (CPM) research project focuses its work on the conservation of built heritage. Additionally, the CPM project works together with the students of the Faculty of Architecture creating a collaborative environment between academia and research. The students go on to occupy an important role in the different initiatives generated by the project and their curriculum is enriched by these experiences.

One of the tools used by the project has been maintenance campaigns. Four campaigns have been launched to promote maintenance in urban and rural sites. The first campaign was carried out in 2011, in the town of Susudel. The second campaign took place in the Susudel cemetery in 2013. These experiences from rural areas were transferred to the city of Cuenca in 2014, in the traditional neighborhood of 'San Roque'. Finally, in 2018 there was an intervention in the neighborhood of 'El Vergel'.

These initiatives focused on preventive conservation are based on the phases foreseen within the preventive conservation cycle, which are: anamnesis (analysis or research), diagnosis, treatment and control (ICOMOS 2003). The anamnesis includes the conceptualization, justification and assessment of the area to be intervened. The diagnosis includes the analysis of the construction state of the buildings and collects general information on the context. The project is promoted and the relationship with the community is strengthened. The treatment encompasses the maintenance process, through the execution of the maintenance campaign. The control evaluates the process and verifies the results of the previous phases. A monitoring system, through short, medium and long-term actions, and the measurement of impact also forms part of the control (Achig et al 2018).

In addition, the participation and support of the community, owners, and public and private companies were essential to generate engagement and work in an environment of many involved parties.

1.3.1 Study case: Street maintenance campaign 'Las Herrerías

The research is based on the maintenance campaign for the "Las Herrerías" street, which is located in a traditional neighborhood of Cuenca, a city declared a World Heritage Site by UNESCO in 1999. The heritage landscape of the place is made up of vernacular constructions of adobe or bahareque, cane roof with mortar of mud, wooden structure and handmade tiles.

The "Las Herrerías" street consists of 77 buildings. The pre-selection of the real estate to be intervened is carried out with the analysis of general information on the facade, the threats and the construction methods, resulting in 44 pre-selected buildings. During the process, we worked together with the owners, identifying their interests in applying a preventive conservation process in their real estate. Finally, 20 buildings were selected and treated. 15 underwent important interventions such as: change of roof, placement of channels for draining water, ceiling repair, wall maintenance (plastering and applying stucco) and wall consolidation, among other maintenance works. The remaining 5 buildings basically had maintenance interventions at the facade level.

During the pre-selection stage, 9 interviews were held and after the completion of the works, 9 more interviews were done of volunteers among the owners, tenants, custodians and neighbors.

In this article we present the thematic analysis of these interviews, the research of which is framed within a socio-critical paradigm focused on meeting the needs of groups of people or individuals. This paradigm seeks a social transformation and aims to change the lives of the participants by enhancing their resources (Ramos 2017).

2 METHODOLOGY

In this research, a qualitative approach has been applied to analyze the impact of maintenance campaigns and the perception of the inhabitants. This approach has been selected since qualitative methods can deeply analyze the experiences, perceptions and opinions of the participants.

Specifically, this research presents the thematic analysis of interviews carried out with the inhabitants of the 'Las Herrerías' street. Thematic analysis is a method for identifying, analyzing, and reporting patterns in data. During thematic analysis the data are not only organized and described, but also interpreted (Boyatzis 1998).

Next, the data collection and analysis processes are described, as well as the tools used in these processes. Section 2.3 describes the self-reflection involved in this research.

2.1 Data collection

The interviews were used as data collection tools. Interviews are widely used in qualitative research to study participants' views, experiences, and beliefs. The types of interviews are: structured, semi-structured and open (Frances et al 2009). In this research, the application of semi-structured interviews has been chosen. In total, 18 interviews of owners, tenants, custodians and neighbors were analyzed. Table 1 shows the statistics of the interviewees.

Table 2 shows the general themes from which the questions in the interviews were generated.

Semi-structured interviews present a list of questions to follow, however new questions can be added.

Table 1. Data of the interviewees.

Interviewee Role		Time of interview application	Use of the building	Intervention on the building
E1	owner	pre-campaign	housing and commerce	Yes
E2	custodian	pre-campaign	housing and commerce	Yes
E3	owner	pre-campaign	housing	Yes
E4	neighbor	pre-campaign	commerce	No
E5	lessee	pre-campaign	housing and commerce	No
E6	custodian	pre-campaign	housing and commerce	Yes
E7	owner	pre-campaign	housing and commerce	Yes
E8	neighbor	pre-campaign	housing	No
E9	neighbor	pre-campaign	commerce	No
E10	owner	post-campaign	housing and commerce	Yes
E11	custodian	post-campaign	housing and commerce	Yes
E12	custodian	post-campaign	housing	Yes
E13	custodian	post-campaign	housing	Yes
E14	neighbor	post-campaign	housing and commerce	No
E15	custodian	post-campaign	housing	Yes
E16	custodian	post-campaign	housing and commerce	Yes
E17	custodian	post-campaign	housing and commerce	Yes
E18	owner	post-campaign	housing and commerce	Yes

Table 2. Themes of the interview questions.

Pre- and post-campaign themes

- Knowledge of the campaign
- Opinion about the campaign
- Benefits of the campaign
- Perception of the heritage values of buildings
- State of conservation and maintenance interventions

Post-campaign themes

- Future recommendations for the execution of the maintenance campaign
- Knowledge of the history of the neighborhood and the building
- Level of participation in the campaign
- Evaluation of the participants involved
- Evaluation of the execution of the campaign for economic, cultural and environmental aspects
- Future maintenance strategies

It is important to emphasize that these initial questions sometimes generated other questions to deepen the topic.

The interviews were applied individually after each participant signed an informed consent form in which confidentiality and anonymity were guaranteed. The audio of the interviews was recorded and later transcribed.

2.2 Data analysis

The data analysis was carried out with the program Atlas.ti, which is a qualitative data analysis (QDA) program that provides useful tools in academic research (Hwang 2008).

The thematic analysis does not require the use of a software program, however Atlas.ti provides a systematized organization since it allows the codification and analysis of the interviews, and the creation of networks

is automated, saving processing time. Additionally, the process in general becomes more reliable since a record of the steps followed is kept, and it allows group work so that extensive projects can be distributed among several researchers.

On the other hand, there are also certain difficulties: the tool is expensive and its free version is limited with respect to the number of quotes that can be made, the project integration process is complex and not very intuitive. However, despite these difficulties, the use of Atlas.ti for this research is considered valuable.

For the thematic analysis of the interviews, the process suggested by Braun & Clarke (2006) was followed, which is described in the following sections.

2.2.1 Familiarization with the data
Of the 18 interviews conducted, 2 pre-campaign and 2 post-campaign ones were taken as examples. These interviews were used for an initial analysis as the interviewees responded extensively to the questions asked, providing a great deal of important information.

2.2.2 Generation of initial codes
The first 4 interviews were used as pilot cases for the generation of codes. Three researchers coded the interviews individually and after a triangulation process an initial list of codes was established.

With this list, at least 2 researchers coded each of the 14 remaining interviews, which were subsequently triangulated. Table 3 shows the list of codes used.

During this triangulation process, some relationships between codes were found that were registered through the use of memos, which were very useful during the automatic construction of networks. The networks were analyzed in depth by three researchers and new relationships were established.

Table 3. List of codes.

Code
Perception of the campaign ($-$)
Perception of the campaign ($+$)
Other needs and suggestions for improvements
Perception about the institutions
Community involvement during the campaign
Perception of the intangible values of the neighborhood
Perception of the intangible values of buildings
Perception of the tangible values of buildings
Perception of the tangible values of the neighborhood
Perception of heritage as a development resource
Perception of the relationship between neighbors
Motivations to intervene
Motivations not to intervene
Tenure
Fragmentation of the property for different uses
Education concerning heritage
* Minga
* Perception of people exogenous to the neighborhood

* Codes added after the triangulation process

Table 4. Initial themes.

Theme
The quality of life, the continuation of families, the uses of the neighborhood and the effects of inheritances.
The perception of the patrimonial values of the neighborhood.
Knowledge of stories, traditions and new knowledge acquired.
General perception of the campaign, of the institutions involved and of other needs other than maintenance actions

2.2.3 Search for themes

With the use of memos, relationships between codes and the final network generated, it was observed that the opinions focused on the following themes:

2.2.4 Review of themes

After an in-depth analysis, the themes were specified in greater detail and organized according to the four pillars of sustainable development defined by Van Balen and Van desande (2015). Table 5 lists the themes and the four pillars of sustainable development: social, economic, cultural and environmental.

2.2.5 Definition and name of themes

The themes were analyzed and described in detail as shown in the results section.

2.2.6 Report production

Some examples of quotes were selected for each theme and the final report was produced.

2.3 Self-reflection

The use of quantitative methodologies can generate interpretations that are forced or aligned with

Table 5. Themes and pillars of sustainable development.

Social
Improving the quality of life
The continuation of families in the neighborhood
Cultural
The perception of the heritage value of buildings and the neighborhood
Education around heritage
Legends, stories and traditions
Economic
Impact of the campaign and the generation of resources for development
Migration of uses in the neighborhood
Inheritances and fragmentation of properties as triggers of conflicts
The perception about the participation of the institutions in the maintenance campaign
Environmental
Other needs other than the maintenance of buildings and their relationship with heritage values
General perception of the campaign

the researchers' beliefs. To reduce the bias that may occur, triangulation processes have been used in both codification and networking.

It is important to emphasize that the results obtained with this study are used to improve the experiences of the maintenance campaigns generating a positive impact on the quality of life and the perception of the actors about the values of heritage. This ideology aligns with the aforementioned socio-critical paradigm.

3 RESULTS

The results of the analysis focus on the social, economic, cultural and environmental aspects. These respond to the research question centered around the impact of the maintenance campaign on the improvement of the quality of life and on the neighbors' perception of the value of their buildings and their neighborhood.

3.1 Social aspect

3.1.1 Improvement of the quality of life

Regarding the social aspect, the interviewees stated that the "Las Herrerias" campaign improved the quality of life in two fundamental ways: livability and a better relationship between neighbors.

Livability: the most important benefits according to interviewee number 12 are health, tranquility and well-being for the whole family, which coincides with several of the other interviewees: "We are feeling well at home, living better and comfortable" (interviewee 13).

The impact of the campaign on the relationship between neighbors: the interviewees stated that the relationship between neighbors in general has been good, with solidarity and mutual help: "the neighborhood is very close, always when there is a program all collaborate, the neighborhood has always been very united" (interviewee 14).

When asked if they think that with the maintenance campaign the relations between neighbors were going to strengthen, the majority responded yes: "I think that with this campaign we are going to be united much more than we already are" (interviewee 1).

During the campaign the support among neighbors was evident and after the campaign the perceptions of the improved relationships between them was confirmed: "All the people of my group, all five of them, we were always united, when we were missing something it...At the end we remained with that support and that friendship" (interviewee 10).

A neighbor even pointed out that he got to know "the neighbor downstairs, who we didn't know before" (interviewee 12).

3.1.2   *The presence of families in the neighborhood*
An interesting fact is that the same families have been present in the neighborhood for many generations. For example, the families Gallegos, Naula, Merchan, etc., who have passed their houses on from generation to generation and currently are still living here. At some point in time it was just these families who were living in this neighborhood: "The whole neighborhood was like one family, because this neighborhood has had an old-fashioned idea of not letting anyone in from outside, they always chased them away, so that they didn't marry anyone from outside, but now..., their parents dead, their grandparents left their inheritance, they sold it, and the neighbors are new, right now we don't know each other" (interviewee 2).

3.2   *Cultural aspect*

3.2.1   *The minga in the neighborhood and its strengthening during the maintenance campaign*
The minga is understood as collective work for a shared goal (Ferraro 2004). An interviewee said that when he was young, there were mingas for the church and the neighborhood. However, some interviewees stated that this collective work no longer exists in the neighborhood and that the campaign didn't awaken it either (interviewees 10 and 11), so it would be considered a lost tradition. A neighbor (interviewee 11) said that "one time" a cleaning minga was organized. As a curious fact, the testimony of a neighbor, which seems to be an isolated case, points out her spirit of collaboration: "As long as I have my spirit of life, my joyful spirit, my spirit of solidarity...and of working with other ideals that benefit the community" (interviewee 1).

During the campaign different levels of collective work were noticed. A neighbor said that he participated in two or three cleaning mingas (interviewee 11), others were organized through directives and were in charge of organizing several aspects of the campaign. Other neighbors participated more actively and with greater affinity, according to the work group they belonged to: "As agreed during the meetings with the entire neighborhood, I personally got involved in giving snacks to the workers" (interviewee 12).

At the end of the campaign, some neighbors emphasized the participation of "all" the owners of the intervened houses: "The inauguration was held, all of us who had their houses intervened participated and we even split ourselves into the same groups as before when we were organizing who was going to do what" (interviewee 10).

3.2.2   *The perception of the heritage value of the buildings and the neighborhood*
At the end of the campaign, a change is noticed in how heritage is being viewed by the neighbors. People are feeling well, they see how nice their houses are, they begin to like the antique, they are satisfied because other people have come to visit them and have learned about traditional construction techniques: "Now people see the repaired roofs, before I didn't even like old things very much, but ever since the campaign began, one could see the change..., I learned how the earth was prepared, how to sift the earth and then use it with a yellow mud and at the end see how it was being applied as stucco" (interviewee 10). "Yes, my house is 200 years old, it is one of the oldest houses. We wanted them to change the whole roof, but they made us understand that since it is part of the historical area, changes are not allowed, since it's heritage, so we haven't touched any of the walls" (interviewee 11). "In a neighboring house it was interesting to see that since the roof had only been tied with cords and this had stood the test of time, it was left intact just as it was before" (interviewee 18).

Several neighbors pointed out that sometimes outsiders value their neighborhood more and they recognize the lack of education about the care of heritage: "Yes because people who come from other places appreciate the neighborhood more...because for example my children don't, they would demolish the house..." (interviewee 10).

3.2.3   *Heritage education*
It should be noted that the majority of people already knew that in order to intervene in their house, the presence of a specialist is required, whom to turn to in case of maintenance (interviewee 13).

With regard to the social aspect, as mentioned before, during generations the neighborhood has been inhabited by the same families. So a transfer of family knowledge and wisdom is common, especially related to the blacksmith profession: "First of all I would like to thank God and my parents who taught me to fight in life. This profession I learned from my ex-husband, it is a very hard profession, since it is considered to be a male profession" (interviewee 14).

3.2.4   *Legends, tales and traditions*
Several interviewees were born in the neighborhood or have lived there for several decades. They shared some

legends that unfortunately are slowly being forgotten: "My older uncles used to talk with my mom and mentioned that over here a carriage passed by that launched fire and which was lost in a ranch that belonged to Miss Florencia Astudillo. It appeared, entered there and got lost, as it was a type of swamp, it was kind of a holiday ranch" (interviewee 2). "What I remember from when I was a child is that the houses weren't as they are today, the street was only an alley, only the horses from Santa Ana, Quinjeo, Jima and San Bartolo passed by, they passed by with loads of firewood and barley for Cuenca" (interviewee 2). Among the stories of the neighborhood, there were those who remembered how the "las Herrerias" street used to be and the importance of some architectural elements, such as portals, which in some sections have disappeared. "What my grandparents, parents and uncles have told me, is that the blacksmiths had houses with portals to tie their horses to, they left their horses to explore the surroundings" (interviewee 16). "This street was for horses, it cost 2 pennies to park the horses at Mr. Moises and Mrs. Juana. There were no cars" (interviewee 11).

3.3   Economic aspect

3.3.1   Impact of the campaign and the generation of development resources

In general, a positive perception is observed about the impact of the maintenance campaign on the economic and cultural aspect. In fact, several neighbors pointed out that through the interventions of the maintenance campaign, the heritage values of the neighborhood, such as its streets and buildings, will be enhanced. This will attract tourists and generate resources for its development, since the neighborhood will be more attractive (interviewee 4). Reinforcing this idea, another neighbor mentioned that the campaign will positively affect the shops, pointing out that "the neighborhood is now more commercial...Now, all that is commercialized here will be sold" (interviewee 7). After the campaign, a neighbor said that "there are shops that are thriving" (interviewee 12).

3.3.2   Changes of activities within the neighborhood

The economy of the neighborhood was based on the blacksmith profession, however in recent years there has been a change to the sale of traditional foods: "There are not many blacksmiths left anymore, but that tradition is still kept alive. Now instead, one can see traditional foods being sold all along the "las Herrerias" street; humitas, tamals and coffee attract lots of people" (interviewee 11).

There are currently two female blacksmiths, which is unusual, since at the same time that the blacksmith profession is slowly disappearing, it is still being performed by women, who are being discriminated, since the blacksmith profession is considered to be a male profession.

3.3.3   Inheritances and fragmentation of properties as triggers for conflicts

A common situation throughout the neighborhood is that many buildings belong to heirs, who more often than not have problems with each other, which hinders the intervention of their properties.

"There are houses that belong to heirs, that are not divided and that we cannot repair, I think that the municipality should help us divide these houses, not with walls, and that each of the owners, well, maintains the houses" (interviewee 1).

"A nephew told me about the house that is already falling apart...He can hardly do anything, since he has to ask permission to his brothers. I am not sure if his brothers will give their permission or not" interviewee 7).

A specific case of a building with two owners was encountered during the campaign. The owners of one part wanted to participate in the campaign, but the owners of the other part didn't, which caused difficulties during the intervention.

The fragmentation of buildings causes the arrival of people that do not belong to the neighborhood, who are often not integrated into the dynamics of the neighborhood nor into its community life, causing insecurity about the people who come to inhabit the neighborhood: "It has changed completely now, it seems that there is crime, they stop here in the park, they stop here on the corner, they stop over there, it scares us because they are strangers, new people, instead before, I remember, 7 or 8 years ago, we were uncles, cousins, nephews" (interviewee 2).

A lot of the buildings are not used by the original people of the neighborhood and are changed from single-family to multi-family housing, affecting the tangible and intangible values of the neighborhood.

"Each of the houses is used to rent out rooms, which makes that other people from other cities come, they give medical students who come to study a chance, they come with another culture, other needs" (interviewee 1).

3.3.4   The perception about the participation of institutions in the maintenance campaign

The procedures and permits that one must request at the municipality cause demotivation to intervene (these include maintenance actions) or have given way to illegal interventions in heritage buildings.

"With regard to permits, for heritage buildings one needs lots of time to go and get permits, you know that after leaving a request they tell you to return after two weeks" (interviewee 1).

"One is not even allowed to paint, since everything is a procedure...the municipality is too bureaucratic, I think, because painting shouldn't be any problem, but they even want to decide which color to use..., I made some changes, but behind the municipality's back, we even installed some zinc roofing sheets, a Friday night we removed the roof tiles, changed the beams and installed the zinc roofing sheets and on top we placed the tiles, so that they wouldn't notice, because if we would have to ask for permission, we would probably still be waiting today" (interviewee 10).

A case in point during the "las Herrerias" campaign is the following: A neighbor wanted to build another floor in the part of its terrace and ask the municipality for permission (without knowing if it was going to be granted). He had the intention to do this during the campaign, but realizing that the campaign was a maintenance campaign, he decided to leave the campaign, after having being selected (interviewee 8).

A question was also raised about the owners' perception about the institutions that participated in the campaign. When inquiring about the responsibility for the maintenance of the buildings, the majority responded that the first to be responsible is the University of Cuenca or the Ministry of Culture and Heritage and the second to be responsible is the municipality. A curious fact is that the owners place themselves last when it comes to the responsibility for the maintenance of their buildings. This shows that in general the owners expect another person or institution to take the initiative to carry out maintenance of their buildings, which should be a priority for them.

3.4 Environmental aspect

3.4.1 Other needs and their relation with heritage values

The campaign shows that there are other needs apart from the maintenance of their buildings that for some neighbors have a priority and that are related to the improvement of the sector and the rescue of the intangible values of the neighborhood. There were owners who mentioned the need to have museums showing the antique handcrafts that several neighbors keep in their houses. Others mentioned that the electric wires should be buried, the sidewalks improved, streetlights installed and the central square to be maintained. To satisfy some of these needs, the neighbors suggested to use the local blacksmiths: "The street should be full of lanterns made by our blacksmiths. And benches so that people who visit us can rest" (interviewee 2).

When asked whether there have been changes to the surroundings of the "las Herrerias" street after the maintenance campaign, some mentioned positive changes: "Yes, especially esthetically, it is prettier" (interviewee 16). However, other neighbors indicated that there were no significant changes to the neighborhood.

3.4.2 Perception of the campaign

The majority of the interviewees expressed a positive perception about the maintenance campaign and indicated that they would participate again. However, some suggestions for future campaigns were raised, especially related to the lack of campaign planning and the coordination of work with the municipality: "They

should be clearer, for example in my case I was told one thing and at the end it wasn't done. Having a better planning and consequently avoiding delays, since I was told that my house was going to be ready in four weeks, but at the end it took them eleven" (interviewee 10).

4 DISCUSSION AND CONCLUSIONS

The criteria that identify the impact of the maintenance campaign on the quality of life and on the perception of the heritage value of the buildings and the neighborhood mainly correspond to the social and cultural aspect. The maintenance campaign contributes in several ways to the improvement of the neighbors' quality of life and the neighbors' view of the value of their cultural heritage. After the campaign several neighbors pointed out that they are living better and that they have learned about traditional construction techniques as well as the heritage value of the neighborhood and its buildings. The relationship between neighbors has improved, they know each other better and collaborate in participatory activities such as the minga. The inhabitants have become aware of the tangible and intangible values of their buildings and their neighborhood, which has led to a better understanding of cultural heritage.

Regarding the cultural aspect, the will to participate through the minga stands out, which increased in a contagious manner during the campaign. In fact, the ones who had never participated in a minga before, mentioned during the interviews that they felt satisfied with their contributions, from the handing out of snacks and the giving of economic contributions to the formation of commissions.

Within the economic aspect, it is observed that demotivation exists to carry out maintenance of the buildings due to the requirements of the municipality. Furthermore, during the campaign there was a lack of agreement between the technicians of the University, the municipality and the owners with regard to the work to be carried out. Additionally, the fragmentation of the buildings and a lack of agreement between owners (heirs) at the moment that maintenance decisions need to be made complicate the interventions. In general, there is a positive perception about the impact of the maintenance campaign. In fact, several neighbors pointed out that due to the interventions of the maintenance campaign the heritage values of their buildings and the neighborhood will be enhanced.

Within the environmental aspect, it was observed that some neighbors did not perceive changes in the surroundings of their neighborhood after the campaign, even though several actions were carried out. The Electric Public Utility Company replaced and relocated light posts, leading to safer circulation and facilitating mobility for the blind, who have an association in the neighborhood. Furthermore, together with several companies, a large part of the aerial wiring was removed to improve the image of the neighborhood. However, based on the reactions of some of the owners, these actions are not always recognized, which is why for future campaigns it is recommended to promote these kind of interventions to raise awareness of the scale of the project and the benefits to the neighborhood. Some owners indicated a lack of planning during the maintenance campaign, especially during the execution phase. This needs to be contrasted with other issues, such as the number of rainy days during the interventions and the delay in the receipt of funds from the municipality. Often, these issues are not properly visualized by the owners, causing annoyance, which should be a lesson for future campaigns.

REFERENCES

Achig Balarezo, M. C., Cardoso Martínez, F., Vázquez Torres, M. L., Jara Avila, D. F., Barsallo Chávez, M. G., Rodas Aviles, T. E., & García Vélcz, G. E. (2017). Campaña de mantenimiento de las edificaciones patrimoniales de San Roque 2013–2014. Universidad de Cuenca. Retrieved from http://dspace.ucuenca.edu.ec/handle/123456789/2872

Achig, M., & Barsallo, G., (2018). "El atlas de daños y su aplicación como herramienta de gestión para el patrimonio de la ciudad de Cuenca – Ecuador". In ASRI: Arte y sociedad. Vol 14, pp. 1–16.

Barsallo-Chávez G (2019), "Plan de conservación preventiva, estudio de caso: la capilla de Susudel". Retrieved from: http://dspace.ucuenca.edu.ec/handle/123456789/32099

Boyatzis, R.E. (1998). Transforming qualitative information: thematic analysis and code development. SAGE Publications.

Braun, V., & Clarke, V. (2006). Using thematic analysis in psychology. In Qualitative Research in Psychology. Vol 3(77).

Cardoso, F; (2012); Manuales de Conservación Preventiva aplicada para sitios arqueológicos Y tramos arquitectónicos Coyoctor, Cojitambo, Chobshi y Todos Santos, Quingeo, Jima; consultoría para el Instituto Nacional de Patrimonio Cultural, Cuenca-Ecuador.

Frances, R. Coughlan, M., Cronin, P. (2009). Interviewing in qualitative research. In International Journal of Therapy and Rehabilitation.

Ferraro, E. 2004. Reciprocidad, don y deuda. Relaciones y formar de intercambio en los Andes ecuatorianos. La comunidad de Pesillo. FLACSO, Sede Ecuador. Ediciones Abya-Yala, Quito Ecuador.

Hwang, S., (2008) Utilizing Qualitative Data Analysis Software a Review of Atlas.ti. In Social Science Computer Review. Volume 26 Number 4.

ICOM-CC- (2008); Terminología para definir la conservación del patrimonio cultural tangible. Resolución adoptada por los miembros del ICOM-CC durante la conferencia trienal, New Delhi, downloaded and translated on julio 20, 2020. Retrieved from. http://www.icom-cc.org/242/about-icom-cc/what-is-conservation/#.UKP7UoecPng

ICOMOS (2003). "Principles for the analysis, conservation and restoration of architectural heritage structures", Zimbabwe. Retrieved from: http://www.icomos.org/charters/structures_sp.pdf

Ramos, C. A. (2017). Los paradigmas de la investigación científica. In Avances En Psicología, 23(1), 9–17.

Stulens, A.2002. "Monument Watch in Flanders: an outline", in Stulens, A. (Ed), First International Monumentenwacht Conference 2000, Amsterdam, 2002:15

Van Balen, K. & Vandesande, A. (2015) Cultural Heritage Counts for Europe - Executive Summary.

Van Balen, K. & Vandesande, A., (2013) Introducción a: "Reflections on Preventive Conservation. Maintenance and Monitoring of Monuments and Sites", by the PRECOM3OS UNESCO Chair. Editors Koenraad Van Balen & Aziliz Vandesande. ACCO. Leuven-Belgium, 2013. ii-vi p.

Vandesande, A., & Van Balen, K., (2016); an operational preventive conservation system based on the monumentwacht model; Conference on Structural Analysis of Historical Constructions, SAHC 2016.

Vermeulen R. (2012) The future of the historic city A study of partnerships in urban heritage rehabilitation in Amsterdam and Vancouver University of Amsterdam, Master Heritage Studies. Retrieved from.: https://www.academia.edu/33585816/

References

Cultural heritage as a source of inspiration for new participatory management approaches

The Future of the Past:
Paths towards Participatory Governance for Cultural Heritage – García et al (eds)
© 2021 Taylor & Francis Group, London, ISBN 978-1-032-02129-4

Inhabiting heritage: The challenges of community participatory management

M. Lacarrieu
CONICET-UBA, Buenos Aires, Argentina

ABSTRACT: The aim of this article is to critique participatory processes in the field of cultural heritage, based on the assumption that tangible heritage has been associated to social distance, while intangible heritage has been related to social proximity and the currency of its expressions for the communities involved. Since cultural heritage has evolved as a field that appears to ignore individuals, but at the same time needs society to resound, is it feasible to promote it as a context for community involvement? Or, is it that participatory processes are a part of intangible heritage, which came into existence because of the need to include groups that had been historically relegated from the field of heritage? Can cultural heritage take on the challenges and their responses through differentiated approaches to participatory action/management?

1 INTRODUCTION

"Ideally, this is the space of the *agora*. This is a utopia that comes from Greek political thought and that refers to the ideal of a space for politics as the freedom of speech and action between equals. It is the symbolic *topos* of the "democratic conversation," to use Arendt's term. Let us note, to think in a decentered way, that this political conversation could also take place under the "palaver tree" of numerous African societies, provided that the village or city also has a place for an "assembly" that is not related to lineage." (Abélès 2003; Mbembe 2010, cited in Agier 2012:14)

The "palaver tree" can be useful to decenter the common and natural ways of thinking (of speaking, too) that we hold about cultural heritage. It can also help us become subjects that express themselves, appropriate spaces, and/or make decisions about the "short term" of their actions in contrast with the "long term" (e.g., being Africans of Bahia, Brazil, during the carnival as opposed to being blacks in everyday life, as pointed out by Agier [2012:19]). Unlike the short term, which concerns the "palaver tree," cultural heritage, considered as the long term, has been filled with silence and marked by the dismissal of social collectives and subjects, who were deprived of their legacies, their vast historical pasts, and even their right to exist among and along with others in the present. This is probably so because, as Candau (2006) suggested, cultural heritage could be referenced to the notion of "anthropology of death"—social groups associated to properties that no longer exist, generations that inherit properties but at some point cease to feel interested in them. (It should be noted that this concept of time is relevant to Western societies, but not to others. Gabriela Eljuri, the opening speaker at the international conference "The future of the past" (Cuenca & Ecuador

2019), points out that time in Quechua and Andean societies is circular, with the past located in front and the future in the back.) As Chastel synthesized, "no single element from heritage is meaningful outside its relationship with the societies it concerns" (cited in Candau 2006:89). Thus, engaging in the field of heritage has proved an unequal, problematic task, colonized by few who were granted legitimacy by the "heritage authority."

This paper aims to critique and reflect upon participatory processes in the field of cultural heritage, based on the assumption that community involvement is closely linked to the emergence and development of the intangible cultural heritage, where the subjects and collectives involved are direct participants in the social expressions they feel identified with. (Tangible heritage came to be associated with non-use and social distance, neglecting theoretical perspectives on its use and value, while intangible heritage became tied to the currency of its expressions for the communities.) From this perspective, and in line with the World Heritage City research project (University of Cuenca), is it possible for cultural heritage to inspire new approaches of participation or participatory management? Inasmuch as cultural heritage has evolved as a field that appears to ignore individuals, but at the same time needs society to resound, is it feasible to promote it as a context for community involvement? Or, is it that participatory processes are nothing more than a new resource for the heritage field, such as those of intangible heritage, which came into existence because of the need to include groups and subjects that had been historically relegated from patrimonialized items, monuments, and buildings? Can cultural heritage take on the challenges and their responses through differentiated approaches to participatory action/management?

DOI 10.1201/9781003182016-6

2 INHABITING HERITAGE: CONTEMPLATION VS. "SOCIAL USES"

"To critically inhabit what we have been forced to inhabit." (Spivak 2003, cited in Rufer 2017:65)

To unravel the terms of inhabiting heritage, the starting point should be the conception of heritage that was established on stories of dispossession, disappropriation, silencing, denial, even of privation, on *loci* of enunciation tied to modernity and colonialism which Rufer (2017:72) described as "a crucial way of creating the world and signifying the present." This implies, according to the prevailing, naturalized view on heritage, that we should begin by dismantling categories of thought, institutional logics, and referential grounds of authority.

The issue of heritage, like the field of culture, was built around this idea of the world in terms of inheritance, cultural externality, and separation from social life, with the intention not to subvert its tangibility or its meaning. But it was never conceived in terms of inhabiting, for this entails, firstly, to integrate it into today's society and, then, to occupy it (Gravari-Barbas 2005). Occupation does not necessarily entail social appropriation. To occupy can be matched to the concept of "reallocating heritage," that is, re-occupying heritage not by changing its form (tangibility), but by transforming its function (Gravari-Barbas 2005: 12-13). It is clear that the notion of occupation continues to lack a subject, or at least it renders the subject invisible as a major character in heritage decisions.

Even when the property—be it a church, a theatre, or a historical house—continues to perform its historical and social functions, the occupation is shaped by the contradictory relationship between possession and dispossession (Gravari-Barbas 2005: 15), but also by the contradictory relations and interactions between the "heritage site" (Lacarrieu 2014) and the subjects involved with those places. However, up to this point, it is only a matter of "users" and "spectators," i.e., attendants to and/or owners of places that are patrimonialized, or visitors/tourists who have a "right to look" at the place, and who generally possess a cultural and symbolic capital that allows them to do so, much like the expert who develops a close relationship with the site which helps them expand, enrich, and commit their viewpoint in a special/specialized way. Laozi summarized these bonds when he said that "the *façade* of a house belongs to the onlooker," and by Victor Hugo when he asserted that "if the use of a building is a matter of its owner, its beauty belongs to everyone" (both cited in Gravari-Barbas 2005:15). These support the idea of contemplation that has dominated the heritage issue.

Does inhabiting heritage solve the problem of subjects being ignored/denied? Yes, but only partially so. Moving towards the idea of inhabiting heritage involves introducing the relationship between the place and the subjects connected to it. As Gravari-Barbas (2005:14) notes, it leads to inhabiting the place, but not necessarily to "being inhabited" by it. The mere relation between a place and its users—at least in the cases where this is possible, since there are places that do not seem to have "owners" or "users" due to distance and time externality—has led experts in the field of heritage to raise the subject of "social uses." Alois Riegl, a 19th-century Austrian author who was ahead of his time, set a distinction between works of art and monuments. He stated that the former relates to aesthetic experience and contemplation while the latter maintains an empathy relationship with citizens. Citizens, he said, assign certain values to monuments since their condition of belonging to patrimony is "implicitly connected to the tangible," but such condition transcends the object as such to include the perspectives from social collectives. It is evident that the significance placed by the author on the "subjective proximity" undermines the idea of "mummification" (i.e., an artifact suspended in time) which, even if much debated, still persisted at the origins of the intangible heritage as a concept (Arjones Fernández 2018).

Nevertheless, actions derived from standpoints such as Riegl's did not materialize until much later. The aggregation of these uses in surveys, catalogs, and file cards, among other tools for patrimonialization, entails the implication of society, not in the abstract—by it consenting that an item is patrimonialized, even when consensus by the people who use the site is distant—but in terms of considering how groups and subjects close to the sites make contemporary uses of them. García Canclini (1999) discussed the social uses of cultural heritage and pointed out the original sin—the notion of heritage was intertwined with several essential categories (e.g., tradition, history, identity, monument, ruins) that are part of a conservationist strategy that called for "heritage to be in solitude" and, therefore, be "pure," decontaminated of any impurities that people could imprint on it (Douglas 1973). This approach to conceiving and acting on heritage has meant and still means losing perspective of the concrete and symbolic disputes around which multiple, varied, and unequal interests from different social groups confer possibilities on heritage, including uses that can alter the item, such as social mobilizations and conflicts that, like was the case in Quito (Ecuador), Santiago (Chile), or El Alto in La Paz (Bolivia), acknowledge historical centers, and the potential for de-patrimonialization/de-monumentalization to challenge the officialized heritage. (The Mapuche's recent actions are an example.)

Encompassing "social uses" is of great importance to create and increase the value of heritage sites, particularly its "use value," which Riegl referred to. Nonetheless, this is possible only in certain places—some have already been emptied of people because they belong to remote pasts, whereas in others, these uses are linked to interactions between the place, the residents, and the different users who share it, and can contribute to its function, although, as we have argued, not to appropriation processes. The "uses" do not account for the social relations and practices deployed by the subjects, which are the base from which they could draw a sense of appropriation of the esteemed places and sites.

It is still difficult to think of heritage as an inspiration to envision models for social participation given that "heritage sites" where social uses are embedded still coexist with sites stuck in the distant past, far from the present and the future of the Western world modernity. We were recently invited by technical experts to a Jesuit *estancia* (rural establishment) in Jesús María whihc is part of the UNESCO World Heritage program along with other Jesuit *estancias* and missions in the provinces of Córdoba and Misiones, Argentina. The motivation was that the local population felt disconnected from the heritage site, which is in fact considered an obstacle that restricts the use of the surroundings, to the point that in the neighboring village of Colonia Caroya, the recent-past immigrant Friulan identity has greater legitimacy than the Jesuit space. This "absence of community" in the different Jesuit *estancias* was and still is related to the fact that a Jesuit identity or a legitimized past is nonexistent in the current community. It is in this sense that the situation was not explored from the perspective of community participatory management.

In sum, "inhabiting heritage" does not only comprise living in a prestigious site endowed and legitimized in the field of local, national, or world heritage, but also to "feel/be inhabited" amid conflicting appropriations of heritage sites.

3 COMMUNITY PARTICIPATION AND INTANGIBLE HERITAGE: FROM EXTRACTIVE ETHNOGRAPHY TO PARTICIPATORY ACTION AND DIALOGIC MANAGEMENT

The above-cited quote by Chastel about the link of heritage to society began to make sense after the adoption of the Convention for the Safeguarding of the Intangible Heritage (2003) by the UNESCO. Until then, there was an idea of the need to understand societies as part of the heritage, but not of them being an inherent part of the processes of patrimonialization nor a reflection on which social groups to get involved. Chastel's words could have answered questions about "inhabiting heritage," particularly about how the representations and daily practices of visitors to and/or residents of a "heritage site" could be affected, inasmuch the values that visitors and/or experts attach to a place do not necessarily match those from the people who live in it.

This first path, which is tightly linked to the "social uses" of heritage, can probably be construed in anthropological terms. For Fabian (1983, cited in Rufer 2017), anthropology became institutionalized around Euro-centered knowledge and the colonial world, and from such position, it spatialized time and also societies, which were unequally distributed in time, in concrete, through denying contemporaneity to others. "But denying contemporaneity of others (Fabian

1983) is also a way of extracting from them their right to coexist from a valid *locus* of enunciation ... and the legitimacy to have a heritage of their own. (Paraphrasing Fanon [2009], the blacks can have neither a true present nor the extensive historical past that the whites do hold...)" (Rufer 2017:69). In sum, the remote, externalized, and legitimate past provided by the patrimonialized items validated and still validates the "absence" of people, who vanished into the past and are unrecognizable in the present. Even though denial and exclusion continue to be part of the heritage value assigned to certain items and places, the adoption of the notions of "social uses" and "heritage inhabiting" have become key in a context of new appearances, demands, and vindications made visible by groups and subjects that used to be marginalized and historically relegated, not only from the field of heritage but also from that world which had been spatialized and temporalized under the direction of colonial power.

To some extent, the idea of intangible heritage came to "correct" these neglects, including the silences attached to heritage items, and shed light on those sectors of societies that had been excluded from the field of heritage and that, in the face of a world of booming diversities, were singled out to be attended, if only through their cultural externalities. But the groups that gained exclusiveness in the field of intangible heritage, mostly in the beginning, were the indigenous peoples—particularly those signaled by their ethnic condition, although not all indigenous or black people were included. The community as a notion and social translation of the groups with which the experts began to establish relationships was key to defining these groups, but above all to interpreting their logic and practices, and to promoting participation.

When it came to intangible heritage, participation was not just encouraged but became downright critical for safeguarding purposes. This type of heritage concerns cultural practices and expressions which are still current in the present. It is in this temporality that experts (known as "facilitators") must live, negotiate, and agree on processes with existing groups and subjects. For that reason, intangible heritage can be said to revert trends about visibility of groups and individuals and community participation. But it is also important to highlight that community participation came into play almost unbidden within the context of patrimonialization processes, in so far as the groups addressed are necessary for the continuity and safeguarding of these expressions.

Cruces (2010:43) has introduced an idea that is worth mentioning here—the "dialogue metaphor" or "dialogue as a metaphor," as he put it. Intangible heritage indeed calls for dialogue, as a mode of discussion, language exchange, and encounters with others, including their bodies, emotions, and sensations. Naturally, there would be no participation if it were not for the presence of subjects constituted in the difference and involved in the heritage issue, that is, recognized by us in that dialogue (Cruces 2010:44). Why does the author refer to it as a "metaphor"? Because while we cannot literally speak with objects/goods, Cruces suggests contemplating other conditions of applicability for concerning landscapes, places, buildings, music, costumes, and other aspects that contribute to molding comprehensive patrimonialization processes. The seeming lack of dialogue with actual people in the case of tangible heritage—although it is also produced based on social groups, as we have argued above—has led to the impression that it is the intangible, or living, heritage that has provided the opportunity for participation in a community context.

However, just as Intangible Cultural Heritage (ICH) introduced the need for community participation, it did not offer new stances. One of the most widespread methods used to promote participation in terms of ICH is that of anthropology—not only by implementing strategies such as interviews, observation records, videos, and photos, but also by placing the anthropologist as an "intercultural translator," particularly anthropologists familiar with certain communities. The patrimonialization of the *kusiwa*, graphic art produced by the Wajapi (Amapá, Brazil) that took the form of a national registry and, then, the UNESCO list, is an interesting example to examine the limitations of this method, considering that the difficulty did not lie with the anthropologist, Dominique Gallois, but rather with the role she was assigned, namely by the National Institute of Historic and Artistic Heritage (IPHAN). The Wajapi, actually, entered the heritage arena after making a request to the anthropologist as the elderly were interested in the recording of their graphic art forms at a time when the young began to attend school, an event that hindered community relations. The major role granted to the anthropologist by the IPHAN derived from her experience with the community because of her previous ethnographic work. This case shows how the method used was and still is linked to the anthropological praxis, in which the anthropologist "extracts," collects, interprets, and translates data for other "non-Indians." (In 2005, Gallois and the Museum of the Indian presented to the IPHAN a joint recommendation to inscribe the *kusiwa* as an ICH for Brazil.) In this sense, if the intricacies of the relationship of the Wajapi and heritage bore relatively positive results, especially some time after their artistic expressions were included in the list, this was not because of the role of the anthropologist as an intermediary and a translator between the community and the IPHAN, but because of the demand made by the very Wajapi. (Upon gaining awareness of the implications of the *kusiwa* becoming ICH, the Wajapi "contested" it by claiming that this artistic expression did not belong to them, but to the communities, that it was sent by the gods, and that they were actually its guardians, thus raising the questions of "property" and "copyright.")

The inventorying of the *milongas* (tango clubs) that we carried out in Buenos Aires, Argentina, in 2013 is another example of how the anthropological perspective is implemented. Our work with 6 informants showed the translation and transcription difficulties that arose from our intention to have the *milongueros* (*milonga* regulars) use techniques learned by anthropologists at university. The issue heightened as the *milongueros* argued that "they knew the *milonga*," meaning that an inventory based on interviews and observation was unnecessary as they were familiar with this social practice on a daily basis and were able to tell it from first hand. The *milongueros* experienced confusion, uncertainty, contradiction, and even discomfort when confronted with the use of certain concepts, of an anthropological viewpoint that was different from that of a *milonguero*, of a bodily dynamic that was alien to tango as a dance, and even of a narrative put in writing common for the facilitators/experts, but not for the *milongueros*. (Note that the previous free and informed consents were eventually given by speech through videos and interviews.) In other words, not only is the approach not new, but it also causes the typical tensions that occur when there is an encounter between an anthropologist and members of a community, whom we usually place on a horizontal plane.

This participatory model is comprehended by people who take part in institutions and the community as generic agents oblivious of hierarchies and power disputes, a context that restricts participation to that undifferentiated community agent called upon to intervene in a process of valorization and safeguarding already initiated by the public authorities and experts. In this sense, "when considering participation, we only think of communities," and how to "pass the power on to them" (Dos Santos Roque 2017), so that we miss the complex map of social actors involved in participatory management.

Participatory Action Research, which draws on Paulo Freire's work, is defended as a method associated with participatory processes in communities that are considered oppressed, and as a strategy that puts the emphasis on social action, people-activated control, and transformations promoted by the communities. Authors who support this approach—which is rarely linked to the participatory processes conducted in the field of ICH—stress elements such as horizontal relations between the experts and the community members, and the superseding of extractivist methods to give way to constructivist perspectives, having the community and its leaders set the agenda in a context of growing empowerment. This participatory action approach does, however, mystify some aspects, including the idea that it serves to build a better understanding of reality in that it is the community that takes the lead as well as the concept of a more horizontal and authentic dialogue (Balcazar 2003).

But, as mentioned above, within the framework of ICH, the interest in the community and the "carriers" of knowledge and practices is greater than the appeal to participatory action as is described above. Probably, as Pérez Ríos (2018) argues, focusing on the "communal language as a decolonial methodological approach" would be a good alternative. The communal language, as the author points out, is "a form of co-construction and sharing of community knowledge between the non-academic indigenous intellectual and other members of the community that is appropriate to transcend and reach the level of academic indigenous intellectuals, because its use could contribute not only to decolonizing knowledge but also to enabling dialogue between the knowledge constructed by the community and academic knowledge" (2018:150). This communal language is a "collegiate activity" that results from discussions among the different, unequal members of the community (even when the grandparents represent the authoritative voice *par excellence*), including the "intellectuals" as referred to by the author. In the view of Pérez Ríos (2018:151), the "grandparents' voices" make communal language distinctive, because how they share their knowledge facilitates the understanding of that language, but also the participation beyond disciplinary canons, protocolized methodologies, and participatory management approaches that come from the external world. The reflections advanced by the author help build the communal language from other resources of the collective and of its common goods, thus erecting it as a "communal didactic strategy" that opposes the replication of methods that are recognized by the states (2018:152).

In the case of ICH, UNESCO and the states, through their institutions, have strengthened participation in the context of the community, but participation itself varies according to the different interpretations and logics of use instilled by the experts, the facilitators, and the managers, rather than by the communities. As a general rule, situations in which no participatory strategies are deployed do not exist anymore, but mechanisms that restrict to accessing information, entering consultations, and making joint decisions between facilitators and communities are commonplace. Joint actions or autonomous actions with support or assistance to independent community initiatives are indeed rare.

Even in the face of such degrees, some contributions made by the intangible heritage perspective are valuable. Firstly, there is community aggregation beyond bureaucratic procedures, which makes the intervention of the community in safeguarding processes possible, although such processes are usually set in motion by the states while the communities are perceived as horizontal, not suffering from tensions or disputes. Secondly, the relevance assigned by some facilitators to social mobilization is seen as a model to build participation and democracy (Toro & Werneck 1996, cited in Vilarino 2014); not only mobilization but also the building of consensus are deemed key by the author. Thirdly, dialogue as an inflection point through which community leaders and carriers are expected to demand rights from and challenge prejudices of the people in charge of public policies, which means dialogue promotes "social empowerment, contributing to the strengthening of self-esteem and identity" (Vilarino 2014:11). Fourthly, free, conscious participation is promoted, even though it is necessary to examine the "management structure," as Dos Santos Roque (cited

in Vilarino 2014:13) notices, to identify the competencies of the people involved in the process, and at the same time "secure inclusion." Lastly, the relevance of decision-making by subjects and groups is emphasized, be it in its extreme form of "citizen power" or in other intermediate forms such as "community deliberation," "social/community control," or alliances with other "control agents."

4   THE CHALLENGES OF PARTICIPATORY MANAGEMENT IN THE FIELD OF CULTURAL HERITAGE: REACHING-OUT STRATEGIES, POLICIES FOR THE COMMON, AND PARTICIPATORY GOVERNANCE

Who is it that participates? What do different actors understand by participation? Is participation a space of consensus? What are the strategies for participation employed by members of the communities, residents of heritage sites, or citizens at large? Is it possible that participation serves as a relevant tool to design public policies for preserving/safeguarding heritage? Does participation promote empowerment, social justice, and community self-regulation?

One development in the field of heritage—particularly the historical and tangible heritage—was the implementation of policies of access to the supply of heritage items, which brought heritage closer to residents, users, and "citizens." The "Night of Museums" is an example. Policies of access are associated with perspectives based on "social uses," or that in which visitors do not get involved with the processes or decisions, but rather approach the items produced by the experts or the expressions developed by the communities only for a short period. Clearly, reaching out to citizens and granting them access to heritage is an attempt to democratize heritage, but it does not imply active participation.

In the fields of culture and cultural heritage, new approaches linked to policies for the common (Barbieri 2014) have been proposed. This interest resides in the fact that the idea of common goods surpasses that of objects/items—which have been key to patrimonialization processes—by bringing in groups from the community with their resources, logics of action, rules, and systems for shared governance. As Barbieri (2014:111) points out, this novel mode of managing participation downplays the prior leading role of the state—an actor that has usually been "harmful, even predatory of many of the resources and the communities"—and poses "collective ways of managing these resources" that go beyond the decisions taken by the state—which still occupies a distinct intermediary role—since it is the communities that determine how to manage spaces (for example, a community museum), thus engendering responsibilities and rules, with institutions only partially intervening. The *minga*—a system for community collaboration where relations of reciprocity and exchange intervene for the accomplishment of common tasks that concludes with a celebration—can probably be examined under this logic. At the University of Cuenca, the *minga* has become a participatory strategy for reaching agreements with the community, for example, to rehabilitate houses with local materials. But as it can be appreciated, the focus is again on the community—an imperative for ICH—ignoring relations of power and inequality, in an attempt to supersede the approach based on reaching out and granting access.

From an ICH perspective, there are novel developments in relation to the community and the inclusion of new actors. An outlook based on decentralizing participatory processes throughout the state but also to other areas, with an emphasis on a participatory approach that is expected to bring balance and become cross-cutting, is starting to prevail. This approach focuses on "shared management," which engages the state and the community as well as independent managers, the academia, the private sector, and society at large. Such "shared management" functions along with "community management," by which the communities involved with the intangible heritage have gained ground. This new approach entails dialogic management not only between the state and the community, or between communities, but also with other stakeholders involved in participatory processes as long as community representation/representativeness is ensured as determined by community members who can perform as "bridges" to connect with other actors, other modes of dialogue, and other patrimonies.

From a more conventional perspective, but also a more comprehensive view of heritage, the participatory governance system is of paradigmatic relevance. This approach intends to supersede both the centrality of the state—which has been key to heritage—and the centering on the community by placing four actors that are deemed fundamental in the heritage process in conditions of equality and parity: the public sector, the private sector, the academia, and society, expecting that the latter will act not as a consumer, visitor, or user, but as an empowered community. The idea of co-producing and co-governing implies a network of relationships, collaborative management, and varying degrees of consensus, all meant to overcome institutional protocols of participation and achieve higher-quality participatory democracy. Still, aligning these actors and overlooking the different amounts of power and resources they have is a challenge, even more so in some heritage contexts. The private actor usually enjoys benefits that the public sector itself provides through exceptionalities and that the community probably finds very difficult to counter, despite the support and assistance it may receive from academic "facilitators." One reason for this is that the idea of the market associated to the private sector does not usually match the economic activity of the subjects connected to a heritage-related activity—for example, crafts and trades, as can be seen in the city of Cuenca, drive the local economy through exchange and circulation dynamics between the countryside and the urban

area—but rather to an actor with economic resources that most likely takes distance from any potential networks that could be used to exchange and share not only knowledge, but also those same resources. On the route of Las Herrerías street, which is part of the World Heritage City Project, the maintenance of heritage buildings is carried out by the public institutions, the community, and the academia—although surely not all of them have the same extent of power. Nevertheless, in what seems to be of most importance for discussing the model of participatory governance, the private actor is the most disengaged, especially when it comes to buildings whose owners seek economic profit, which heritage can offer in terms of symbolic valorization, which can also constrain material capitalization. There may be other cases, such as one that occurred in Buenos Aires, where a collective of neighbors took on a claim to demand the patrimonialization of houses with some historical relevance, disputing power and producing "gate-crashing participation," which enabled the neighborhood association to force a decision although power had not been explicitly delegated to it. Nonetheless, even after the decision to create a historical protection area was taken, private owners activated strategies and resources so that the houses only had to conserve their *façades*, while other measures the neighborhood collective did not want were taken in secret. Co-management implemented with a participatory perspective can be a promising road for the field of heritage, but it can also be extremely complex, depending on who can activate the power.

Decolonial scholars, like Walter Mignolo, refer to the disobedience of subjects, conceived as a form of questioning structures, but also of overcoming the universalized authority that these authors associate with the coloniality of power. In this perspective, disobedience is not just a spasmodic reactive act that, as Rufer (2017:76) claims, the powerful see with contempt, but actually a process that could end up undermining power and authority. Indeed, it seems unlikely that cultural heritage will escape the conservationist and institutional character that has forged it to embrace or prompt disobedient participatory processes. Even in the case of intangible heritage, where models seem more flexible, prevailing communitarian strategies tend to be de-politicized. Consequently, just as Agier (2012) posits the decentering of the subject, i.e., a subject that expresses itself, takes initiative, and occupies spaces, it is plausible to speculate that in the field of heritage, this very same will only become possible if patrimonialization itself is decentered.

REFERENCES

Agier, M. 2012. Pensar el sujeto, descentrar la antropología. *Cuadernos de Antropología Social* 35: 9–27. Buenos Aires: FFyL, University of Buenos Aires.

Arjones Fernández, A. 2018. Aclaraciones conceptuales y valorativas de la teoría de Alois Riegl para la conservación de pinturas murales. *Conserva* 23: 13–24.

Balcazar, F. 2003. Investigación acción participativa (IAP). Aspectos conceptuales y dificultades de implementación. *Fundamentos de Humanidades* IV(7–8): 59–77. San Luis: University of San Luis.

Barbieri, N. 2014. Cultura, políticas públicas y bienes comunes: hacia unas políticas de lo cultural. *Kult-ur. Revista interdisciplinària sobre la cultura de la ciutat. Governança de la ciutat i drets culturals* 1(1): 101–119.

Candau, J. 2006. *Antropología de la memoria.* Buenos Aires: Editorial Nueva Visión.

Cruces, F. 2010. Sobre el diálogo como metáfora del patrimonio cultural. In E. Nivón & A. Rosas Mantecón (coord.), *Gestionar el patrimonio en tiempos de globalización.* Mexico City: UAM Unidad Iztapalapa & Juan Pablos Editor.

Dos Santos Roque, L. 2017. Participación y planes de salvaguardia: reflexiones, desafíos y perspectivas. *Expressao socioambiental, 10 June 2017.*

Douglas, M. 1973. *Pureza y peligro. Un análisis de los conceptos de contaminación y tabú.* Madrid: Siglo XXI.

Gallois, D. T. 2005. Os Wajapi em frente da sua cultura. *Revista do Patrimonio Histórico e Artístico Nacional,* 32/2005: 1–14.

García Canclini, N. 1999. Los usos sociales del Patrimonio Cultural. In E. Aguilar Criado (coord.), *Patrimonio etnológico. Nuevas perspectivas de estudio, Cuadernos* X: 16–33. Granada: Instituto Andaluz del Patrimonio Histórico.

Gravari Barbas, M. 2005. Introduction Générale. In Gravari Barbas, María (dir.), *Habiter le patrimoine. Enjeux-approches-vécu.* France: Presses Universitaires de Rennes (PUR).

Lacarrieu, M. 2014. "Patrimonios vivos" en tensión: entre la fijación y la movilidad se "hace patrimonio" (unpubl.)

Pérez Ríos, E. 2018. El lenguaje comunal como aproximación metodológica decolonial. In AVÁ (33) *December 2018.*

Rufer, M. 2017. Temporalidad, sujeción, desobediencia: de algunas premisas de Walter Mignolo hacia una crítica para pensar históricamente. *Epistemologias do Sul* 1(1): 60–86.

Vilarino, M. 2014. La participación de la comunidad en la gestión del PCI. *Formación para la gestión del patrimonio cultural inmaterial en el ámbito de la COOP SUR. Curso libre a distancia (EAD).* Brazil: Inspire.

The Future of the Past:
Paths towards Participatory Governance for Cultural Heritage – García (eds)
© 2021 Taylor & Francis Group, London, ISBN 978-1-032-02129-4

Reflections about the sense of cultural heritage. Case of Cuenca, Ecuador

F. Cardoso & G. García
City Preservation Management Project, Faculty of Architecture, University of Cuenca, Cuenca, Ecuador

ABSTRACT: On the occasion of the second century of the independence of the city of Cuenca, this contribution aims to reflect on the dynamic process of becoming a World Heritage Site, and some of the key actors involved in that recognition process. In chronological order, the first section describes the socio-cultural and physical environment at the national and local scale in the 18th and 19th century, based on relevant descriptions written by foreign visitors. The second section identifies some events, considered as milestones in the transition of Cuenca from a colonial city towards a modern city, emphasizing the global debate between transformation and conservation during the 20th century. The third section refers to the key actors in the valorization of Cuenca as heritage, initially as national heritage (1982) and later as World Heritage Site (1999). It concludes highlighting the "mestizaje" as an essential aspect of its outstanding universal values.

1 ECUADORIAN PHYSICAL AND SOCIO CULTURAL ENVIRONMENT

For the present analysis some historical descriptions of the Ecuadorian and local context are cited. On the one hand, Simón Bolivar "The Liberator", an enlightened man, with knowledge of the world, wrote around the beginning of the 19th century:

"I arrived here a few days ago, having been well received and magnificently gifted. The people seem to be good, although not all of them say the same thing. The country seems miserable because it lacks everything, except grain which is in abundance but without means of transport. Here, the cleric is everything, while the Indians are nothing because they are poor and few".

In Bolivar's expressions, a sketch of the mentality carved out by centuries of dependence and colonial domination of this society, also comes to light. After leaving Guayaquil where he made friends with the Garaicoa family, Bolivar wrote:

"...To the Glorious One, that I have liked the Serranas (women from the highlands) very much, although I have not seen them yet; may she not envy them as she said, because she has no cause with such modest people who hide from the presence of the first military man. The church has taken over. I live in an oratory. The nuns send me food; the canons give me refreshments. Meditating on the beauties of providence endowed to Guayaquil and on the modesty of the serranas, who do not want to see anyone for fear of sin. In short, my friends, my life is all spiritual and when you see me again, I will be all angelic".

On the other hand, some years before, Laportdi (1797) refer to Cuenca as follows: "*The city can be considered fourth-order in its extension. Its streets are straight and quite wide, the houses are made of adobe, covered with tiles. Those in the neighborhoods or suburbs are untidy and rustic because they are the ones inhabited by the Indians. In the middle of the city, they cross several ditches taken from the rivers...*"

Both descriptions agree that Ecuador and Cuenca were territories of great beauty and enormous poverty. Additionally, some of these expressions also show crudely the reality of the society of that time: a society constituted by a dominating class, which lived from the exploitation of decimated human groups, with reduced rights, and socially marginalized. In the best of cases, they were considered free or very cheap labor for tilling the soil, for construction works, or for domestic services.

It means, in addition to the obvious precariousness of infrastructure, services, and productive capacity, there was also – or as the consequence of – the existence of a pyramidal social structure. Indeed, 250 years after the conquest began, the features, roles, and personality of society were clearly delineated: an elite that had consolidated power, taking possession in the region of the best lands (including Indians) through the resource of the "encomiendas", a growing class of Creole mestizos, and a social and productive basis based on the enslaving exploitation of the regional indigenous communities. The work of these last ones, fundamentally linked to agriculture, but also to mining, to the construction of the always precarious infrastructure or to the production of textiles, was free or very badly paid, and many times paid in species or favors, which only gave the people of the countryside, a condition of precarious survival.

The Church joined the repressive role of the executive power, expressed in the governorships of the Royal Audience, with little shame. Following its interests,

DOI 10.1201/9781003182016-7

the ecclesiastical power accumulated lands and properties of great wealth, which above all contributed to the mental and ideological colonization of the local indigenous people. In fact, at the beginning of the 19th century, the persistence of a conservative society, obsessed with the idea of sin, is evident in Bolivar's description in his letter to the Garaicoa sisters of Guayaquil. That reality was the result of the convergence of tools of social, military, ideological, and spiritual domination, , which consolidated the colonial society, whose characteristics were also expressed in the urban world, and in the society of the small Cuenca of 200 years ago.

Departing from those scenarios, we should ask ourselves two fundamental questions about the period that separates us from the moment of independence: What happened in the following 200 years with the city of Cuenca to consider it -and to succeed- in being included in the list of the world's cultural heritage? And – looking to the future – what does this heritage mean for a contemporary society like ours?

2 FROM THE MISERABLE CITY TOWARDS THE HERITAGE CITY

In the following one hundred years, around 1920, Cuenca worked to become a society that wakes up slowly, feels the need for self-determination, and builds its path to the future. An example of this was the ephemeral existence of the Republic of Cuenca, which was more symbolic than real. It initiated with slow, tortuous, exhausting efforts to take firm steps towards the construction of a more educated society, with the creation of schools and universities, productive and craft associations, social guilds, printed and electronic media.

To this was reinforced by the proactive attitude of the elites and of some religious congregations, striving to modernize the city, providing infrastructure, energizing the economy, building great monuments – influenced by those of the metropolises with which its commerce was related in Europe – and renewing buildings inherited from the colonial period. All of this in the search for a social expression of the city of the future.

At the beginning of the 20th century, Cuenca experienced important transformations. In the context of the commemoration of the first centenary of independence (1920), squares, streets, private buildings, electric and water infrastructure, churches, roads, bridges, and even an airport, were built.

The colonial world was gradually fading, at least in its formal expression. It opened up to make room for a universe full of eclecticism and visions of modernity, with symbols as powerful as the arrival from the coast of the first car in Cuenca, loaded on shoulders, or the landing of the daring flight of Elía Liut and Ferruccio Ricciardi, from Guayaquil.

The lack of accessible roads to connect Cuenca with other cities due to the mountanous topography implied the use of old chaquiñanes (prehispanic road system).

Figure 1. Cuenca in a future of 50 years. Plan regulador de Cuenca Arch. Gilberto Gatto Sobral 1946.

People – indigenous mainly- crossing wild territories to bring technology or sumptuary items such as mirrors, pianos, brass to beautify the spaces with a halo of european dignity, that was conceived in the stately homes of Cuenca. This lack also influenced the provision of infrastructure in Cuenca that was epic. As an example, the generators for the first electric plant were brought on the shoulders of indigenous people by around 150 km, from Huigra to Cuenca. Recently in 1965 Cuenca joined the rest of the country through the railway.

Another important event in the process of transformation of Cuenca came with the first urban regulation plan, in the mid 20th century. It implied establishing extension areas for the consolidation of the city of the future, that of the monumental buildings, of the automobile mobility, of the great avenues, and the garden city. The Faculty of Architecture, created in 1958, played a crucial role in implementing locally the postulates of the international modern movement. In this way, the foundations of the New City were created, similarly as 400 years ago, when conquers found an ample and clear space to design the city of the checkerboard, modernism found in the called Ejidos, the place to promote the city of the future.

In that way, Cuenca co-existed between the legacies of the past and new ideas for the future. A breath of freshness and innovation was added to the old checkerboard layout inherited from the colony. The balconies of colonial buildings with large walls barely open to the public left room for a rhythmic, vibrant, innovative architecture in which the mestizo spirit of society was not lacking. It was a society that presented a renewed thought regarding the valuation of heritage where the architecture described as miserable in 1800, gradually became endearing architecture 150 years later.

3 HERITAGE: TENSIONS BETWEEN CONSERVATION AND MODERNITY

But, in addition, the aforementioned transformation of the old colonial city into a new one, came with countless replacements and demolitions. The transformations involved not only private properties but also public and ecclesiastical ones, because it was seen as a natural fact that the updating of the aesthetic taste or the functionality of the buildings, was adjusted to the contemporary needs and aspirations.

One of the buildings that entered in the list of demolition projects was the old Government House, Prison, and Barracks in Cuenca. Photographs of it are preserved, as its use had remained almost unchanged from the very moment of the colonial foundation in the central square of Cuenca. The government house was to make way for a modern public service building. The threat of its demolition aroused a well-argued defense of its conservation, what could be considered the oldest expression found in the history of the conservation of the city's heritage up to now. This defense arose not so much for practical or utilitarian reasons, and not even for a certain monumental value since this colonial building was a modest work, but rather for having historical attributes and memory, kept in its walls and in the facts that this century-old building maintained.

The task of protecting the material legacy from the past was not easy. There was an idea of modernity installed in the city's elites (and also in the local authorities), with a discourse that created the false dichotomy between conservation and progress. At that time, heritage conservation started from an attitude – almost desperate – of opposing the demolitions and replacements that were taking place in every part of Cuenca.

However, that could be considered the early precedent of what would happen later, especially around 1970, when citizen initiatives promoted the first organic and sustained process that focused on protecting the city's cultural assets. As consequence of that, in 1982, Cuenca was listed as a national heritage that allowed to move towards deeper and more decisive actions in favor of heritage protection.

The idea of a heritage city, stimulated a process of specialized training for professionals and academics. In 1988, the University of Cuenca, through its Faculty of Architecture, promoted the first postgraduate course called "Specialization course on the conservation of monuments and sites." It was the response of

the academic sector, to provide better and more sensitive intervention practices. Since then, the Faculty of Architecture has incorporated conservation plans initially in the chairs of Urbanism, and later on, it created a new training area for students referring to the conservation of built heritage.

In this way, the University established the political and technical foundation for what could be considered an utopia: Cuenca included in the world heritage list. From the political point of view, the impulse was given by a former dean of the Faculty of Architecture of the University of Cuenca (FAUC), elected mayor of the city in 1996, and from the technical point of view, some professors of the same Faculty worked to face the challenging task of building a contextualized and solid argument to propose Cuenca's candidacy to UNESCO. The file was ready in June 1998 and on December 1, 1999, in Marrakech-Morocco, UNESCO, through the World Heritage Center, announced that the historical center of Cuenca became part of the heritage of humanity.

3.1 *The Heritage City 20 years later*

Since 1972 UNESCO, through its Convention Concerning the Protection of the World Cultural and Natural Heritage, aims to identify and protect the world's natural and cultural heritage considered to be of outstanding universal value. Already for 1998, UNESCO had developed specific criteria for nominations, and those criteria (in a differentiated form for cultural and natural heritage) were to be made explicit through values.

Values should be understood as the reasoned identification of heritage aspects and dimensions that, based on credible and documented sources, are capable of showing the universal exceptionality of a monument or site. In the file of Cuenca, in addition to the presentation of a reliable narrative of the heritage site, strengthened by consistent documentary support (graphic, photographic, historical, and management), the identification of its exceptional universal values became the essence of Cuenca's approach. The identification of these values meant, at the time, an exercise aimed at clearly deciphering the dimensions applicable to Cuenca, in its national, Latin American, and universal context.

According to UNESCO, the Historical Center of Santa Ana de Los Ríos de Cuenca was inscribed on the World Heritage List for the following criteria: "Criterion (ii). be the manifestation of a considerable exchange of human values during a certain period or in a specific cultural area, in the development of architecture, monumental arts, urban planning, or landscape design. Cuenca illustrates the perfect implementation of Renaissance urban planning principles in the Americas. Criterion (iv): be an outstanding example of a type of building or architectural or technological ensemble or landscape that illustrates a significant stage or stages in human history. The successful fusion of the different societies and cultures of Latin America

Figure 2.   Colonial Goverment house. n/d.

is symbolized surprisingly by the layout and urban landscape of Cuenca. Criterion (v): constitute an outstanding example of traditional human habitat or establishment or land use, which is representative of a culture or cultures, especially if they have become vulnerable due to the effects of irreversible changes. Cuenca is an outstanding example of a planned Spanish colonial town in the interior lands". But what is the profound meaning that this heritage holds for our society and humanity?

If criterion (ii) is considered, it calls us to understand Cuenca as a concrete work in which the urbanistic thought defined in the European Renaissance, fits in the cities of Spanish foundation. A notion of the open compass, with the idea of growth ad infinitum, at least from the theoretical point of view, which could hardly be applied in the old European medieval structures, dominant in the 16th century, and which of course, responded to other logics and other historical processes, consolidated in Europe. Cuenca is not a unique case, but it is one of the most successful and best-preserved, so its degree of universal representativeness is very high.

Criterion (v) refers to the rich urban history of Latin America. The colonial city that consciously takes its distance from the pre-existing nucleus in Pumapungo and Tomebamba, to smoothly locate the colonial settlement with a clear idea of the future in a site of exceptional beauty, differs from the seafaring city, the fortress city, the mining city, the river city and other typologies that unfold in the enormous continent. Its condition of city is clear, it includes indigenous communities, -although in a segregated way- and its process of consolidation along the centuries is slow (almost 4 centuries under invariable guidelines), but very precise in its spatial conformation. Hence, the management of this urban value that is inserted in the line of Renaissance urban creation, requires a concern for public space and for the elements that constitute it, among them local architecture, its uses, its openness and inclusion, its security, the degree of integration that it raises, that is to say, an organic understanding which is closely related to its social uses.

Concerning criterion (iv), the successful fusion of the different societies and cultures of Latin America is symbolized surprisingly by the layout and urban landscape of Cuenca. This feature is particularly exciting, since the mestizaje, which is essentially the origin of a new culture, is the result of a barely harmonious, but symbiotic coexistence. A coexistence marked by inequality and by the exploitation by one sector (small, minority, powerful) of another (pre-existing, majority, weakened by the new conditions of domination). From this coexistence the current society is born. The file of Cuenca indicates that "the mestizaje (has) become a new reality for the American towns, the one that, in the case of Cuenca, assumes urban forms when foreseeing from the beginning spaces of coexistence between Indians and Spaniards within the same citizen territory: the sequence of squares in direction east west explains it with simplicity: compatible models of life,

diverse cultures: San Blas-Indígenas, Mother Church (future cathedral) and central square – Spaniards-San Sebastián, Indígenas.

The outstanding values of Cuenca's architecture not lie so much in the monumentality of its buildings, rather in its unique capacity to adapt itself to the various architectural trends of the past. An adaptation that is reached preserving its essence as a colonial city, which maintains its maximum support in the schemes of its monasteries and civil architecture. While technology and spatial design are the fruit of ancestral Indian and European knowledge, the expressive form chooses the models of European architecture and adapts them to the conditions of the environment.

The message that emerges from these segments of the Dossier of Cuenca, underlines the need to consider heritage assets adopting an integrated perspective for their conservation, especially including the more modest ones, since each building preserves a history, a history of life, taste, and wisdom expressed in technology.

Furthermore, in the case of Cuenca, the mestizaje is also evident in the most intimate and sublime expressions of culture and spirituality of the city. this is evident, until today, not only in its architecture, but also – for example – in the colonial mural painting of the Refectory of the Monastery of El Carmen de la Asunción, in the enormous popular festival of the Pass of the Jesus Child on December 24, or in the dishes offered today by the thriving sector of Cuenca's gastronomy, well prepared to take up the challenge.

Mestizaje, for this heritage city, is a fundamental issue. The social demonstrations and the actions of peasant groups, such as the recent ones in October 2019, are an invitation to rethink the heritage not only from a monumental, architectural or urban perspective but also from a social perspective. Cuenca was built with the commitment of the elites, but above all with the enormous effort of wise local builders, blacksmiths, carpenters, mestizo and indigenous artists, stonemasons. Colonial or republican Cuenca is a city fed by the products of the countryside, procured by wise farmers who defined their cosmovision with the production of the land.

For the conclusion of this section, it is relevant to quote Isabel Allende, the well-known Chilean writer, who recently received an award for her literary work in Spain. She said in reference to the current world: "We

Figure 3. Mural paintings of refectorio del Monasterio Colonial de Las Conceptas de Cuenca, S.XVIII. Source: Fausto Cardoso, 2016.

live in dark times... history, with capital letters, is written by the victors, almost always the white men, who carve their names in marble (...). The dispossessed of the earth are silenced. These are the voices that whisper to me and persecute me and tell me their lives in the silence of the dawn."

These phrases are an invitation to reflection. Or is it not the result of a long history of injustice and marginalization that is lighting fires in Latin America?

The reproduction of inequity in Latin America, the origins of which can be better understood thanks to what heritage discourse offers us, must be studied and analyzed permanently to question, combat and decrease it, as quickly as possible. This is a fundamental role, not only cultural but social that it is up to it to assume the field of cultural heritage.

4 CONCLUSIONS

For Cuenca, the honorable condition as World Heritage Site has imposed great challenges. The city and its heritage are obliged to be rethought permanently, its values to be discussed, questioned and recreated in the light of new approaches and new actors. From the idea of a heritage city for tourists, we must move on to a heritage city for local people (and for tourists). Preserving vitality is one of the great challenges for heritage cities. And vitality means local inhabitants, expressions of culture, places for leisure and social enjoyment, spaces for creativity, opportunities for innovation, but fundamentally spaces for conciliation and social respect.

The mestizaje is a historical fact, which results in the constitution of a new society. Such a society expresses itself in different ways: in the urban organization, in the technology of the oldest buildings in the city, or simply in the way in which even in ages closer to ours, the local culture assumes the language of society and its dominant elites in its architecture. This was also the case for Cuenca, where throughout the 20th century, architectural transformations were influenced by this last phenomenon.

The powerful idea of mestizaje, which was highlighted as one of the outstanding universal values of Cuenca, invites to take advantage of a modest heritage to strengthen the sense of community. In that perspective, heritage becomes a clear resource for the development of people that goes beyond the economic dimension often related to touristic activities, to be a powerful tool for sharing wisdom, interconnection, appreciation, and therefore, respect for diversity.

It has taken thousands of years for the world's societies to become aware of the fragility of their host environment. The globalized world has allowed us to feel that we are part of a whole, in which the pulsations on one side, however small, have repercussions on the other, as the metaphor of the butterflies fluttering illustrates well. Everything is somehow interconnected. Social mobilizations – in diverse and distant points on the planet and particularly in America – are eloquent expressions that something is not right. It is necessary to think of committed solutions that include the marginalized (nations, peoples, communities) because partial balances continue to be imbalances, islands of ephemeral wellbeing that can succumb to new and unexpected circumstances, in the face of which societies have not yet built consistent and sustainable responses.

From the past, we have received a world of extreme beauty and great wealth, which is constantly changing. In this process, the human being has been inserted for only a few millennia, not only with wonderful works of architecture and art, but also with ingenious ways of assuming the natural conditions of their places of life. He has built close relationships with nature that today we consider to be expressions of culture. That is to say, not only the culture that is witnessed by the well-placed stone, the molded earth, the carved wood, or the transformed space for the habitat, but also the accumulation of wisdom built from experimentation, risk, rite, memory, the heritage turned into ancestral wisdom and science, true foundations of contemporary culture.

From this perspective, the future of the past is also the future of the present, and the future of the future, of the new generations that will come and for whom we must commit ourselves to leave them a world that should be better than the one we received.

REFERENCES

Albornoz, B. 2008. Planos e imágenes de Cuenca, Fundación Barranco. Cuenca.

Cardoso, F. Pauta, F, Sánchez, M, & Jaramillo, C. 1998. Propuesta de inscripción de Santa Ana de los ríos de Cuenca en la lista del patrimonio de la humanidad. Municipalidad de Cuenca.

Cardoso, F. 2010. Expediente retrospectivo de la Declaratoria del Centro Histórico de Cuenca en la Lista del Patrimonio de la Humanidad. Municipalidad de Cuenca.

Cardoso, F. 2017. Propuesta de inscripción del Centro Histórico de Cuenca Ecuador en la lista del patrimonio mundial. Universidad de Cuenca/GAD Municipal del cantón de Cuenca.

León, L. 1983. Compilación de Crónicas, Relatos y descripciones de Cuenca y su Provincia, Banco Central del Ecuador. El viagero Universal o noticia del mundo antiguo y nuevo Tomo XIII Carta CLXXXVI pp 150 a 154 Villalpando Madrid, 1797.

Bolívar, S. 1947. Carta a Francisco de Paula Santander, in Obras completas de Bolívar, Tomo I Edit. Lex, La Habana, Cuba.

Municipio de Cuenca. 2010. Elia Liut. El Vencedor de los Andes 90 aniversario de I vuelo transandino Guayaquil-Cuenca. Biblioteca Municipal de Cuenca.

UNESCO. 1972. Convención sobre la Protección del Patrimonio Mundial, Cultural y Natural, https://whc.unesco.org/archive/convention-es.pdf.

The Future of the Past:
Paths towards Participatory Governance for Cultural Heritage – García et al (eds)
© 2021 Taylor & Francis Group, London, ISBN 978-1-032-02129-4

Heritage and cultural management: Participatory experience in Cuenca

A. Astudillo
Faculty of Social Sciences, KU Leuven, Leuven, Belgium

E. Acurio
Faculty of Design, Architecture and Art, Universidad del Azuay, Cuenca, Ecuador

ABSTRACT: This article reflects on the link between cultural and heritage management and civil society participation, through the experience of Agenda Ciudadana. Within this frame, open citizen forums were used as participatory methodologies to identify problems and co-construct solutions with multi-actor stakeholders. As a result, it offers a participatory management model, an action-oriented model, dealing with real-world problems, and enhancing decision-making from civil society. Furthermore, this participatory process brought up heritage management as a discussion point for citizenship. The article lays out the central citizen collective Cuenca Ciudad para Vivir concepts related to heritage and cultural management. The second section explains the methodological approach used in the forums held for Agenda Ciudadana. The third section delineates the problems identified and the possible solutions for them. To end up with a contextualization of the heritage and cultural management during the recession and health care crisis due to the Covid-19 pandemic.

1 INTRODUCTION

In the Latin American context, the management of heritage and culture must face several challenges and adapt to each territory (Timothy & Nyaupane 2009). At the regional level, the cultural landscape is constituted by historical power-relations and tensions deeply rooted in its postcolonial condition. As a result, the subaltern identities contest the official national histories and the traditional comprehensions of heritage and culture (López Caballero 2008; Williams 2002). In this line, Kaltmeier & Rufer (2016:3) point out that *'While in Western European countries, heritage has been transformed into a depoliticized lifestyle factor, heritage in postcolonial contexts has become a battleground on the interpretation of history and its projection into the future'.* Since Latin America's cultural matrix is settled in the encounters and discounter of diverse worldviews, García Canclini (1989:71) suggests that it is the result *of "the sedimentation, juxtaposition, and intertwining of indigenous traditions, Catholic colonial Hispanicism, and of modern political, educational and communicational actions (...) Besides, interclass miscegenation has generated hybrid formations in all social strata.*

In addition to the historical and cultural context, South America is the most urbanized (ONU-Habitat 2010) and unequal region in the world (CEPAL 2016). Urban growth radically changed the economic and socio-cultural reality, increasing segregation, gentrification, and environmental impacts (Pérez & Tenze 2018). Therefore, heritage management must be included in sustainable urban planning that addresses rapid urban growth and the gaps it generates (Hosagrahar 2018).

This article grounds its reflections on heritage and cultural management in Cuenca's particular experience; it is located in Ecuador's southern Andean highlands. This city has several historical layers shaped by different cultures. Originally, it was the indigenous-cañari central settlement 'Guaponledic.' Later on, the Incas occupied 'Guapondelic,' establishing the city of 'Tumipampa' (Burgos Guevara 2003; Idrovo 1985). Finally, the Spanish colonial city of Cuenca was founded in A.D. 1557 (Jamieson 2008). Currently, Cuenca has an approximate population of 636,996 inhabitants in 2020 (INEC 2017). As an intermediate city, it still presents the possibility of proposing other forms of management (Hermida et al. 2015), projecting itself towards a more sustainable and socially-environmentally fair future (Bolay & Rabinovich 2004). Besides, Cuenca holds a World Heritage Site declaration, settling heritage and culture as strategic areas in urban planning and awakening the citizen interest and engagement.

In this context, urban management seems challenged by global dynamics like climate change, international markets, massive tourism, and migration (Pérez & Tenze 2018) and local aspects such as recession, urban planning, unemployment, essential services provision, heritage safeguarding, and cultural management (CEPAL 2016). Furthermore, all these regional and global aspects affect the built environment as the landscape, mobility, housing, built heritage. It also influences the city's intangible dimension like the intangible heritage management model, security perception, sense of belonging, identity, art production, culture, and lifestyle (Borja & Castells 2013).

In the face of challenges, urban management must have a holistic approach to the territory, addressing the local demands within the global context (Guzmán et al. 2017; Irazábal 2017). But also, it must be able to articulate social-political, economic, cultural, and infrastructure dimensions to develop context-based solutions (Borja & Castells 2013). Citizen participatory experiences in Cuenca and the particular experience of Agenda Ciudadana[1] (CCCV 2019) demand a sustainable territorial management model that effectively incorporates values like diversity, social and environmental justice, and equity grounded on greater public participation in decision making and new forms of democratic practice (Asmal 2016; Molina 2019; Pérez & Tenze 2018). Consequently, citizen participation turns to be strategic to build the city, especially heritage and culture, as a collective project by its inhabitants, rather than the particular interest of elites.

1.1 *Citizen collective Cuenca Ciudad para Vivir*

In this line, since 2010, the citizen collective Cuenca Ciudad para Vivir (CCCV) stands as a social organization focused on the reflection and construction of public ethics (Cortina 2008; Michelini 2007). The CCCV understands public ethics from the historical, cultural, political, and social contexts of local realities, guided by the principle of '*Corresponsabilidad Solidaria*[2]', through broad and inclusive participation (Michelini 2007:39). In line with Lefebvre's (1991) contributions, public ethics resides in daily life, the relationship of coexistence and care established in cities as an open and heterogeneous community. Hence building public ethics upon citizen participation and experience constitutes an authentic social revolution (Taylor 1997). In this frame, the CCCV advocates for a change in the citizen's political culture through opening up spaces of dialogue with different actors and strengthening active participation in public decision-making (CCCV 2012). CCCV promotes button-up civil organization in Cuenca in line with theoretical-political proposals such as the Right to the City (Harvey 2008), Participatory Methodologies (Villasante 2011; Villasante 2006) and Participatory Action Research (Borda 2009).

The Ecuadorian Constitution assumes public participation as a right, and since 2010, COOTAD[3] states that the Municipalities must establish systems for citizen participation. Despite this, in the last ten years, parties and government models face a lack of legitimacy and credibility, given the weak community-based strategies implemented (Braga & Casalecchi 2019; Mainwaring & Bejarano 2008). The result of the citizen perception survey held in Cuenca by CCCV shows that: 69% of people have not heard about participatory public management, 88% stated that they had not participated in this type of process, and 56% indicated that they did not meet with their neighbors or others to discuss public problems (CCCV 2015) (CCCV 2016). In this context, CCCV acknowledges the wear of government-driven participation. Therefore, CCCV comprehends participatory democracy as an organizational ethos and praxis more than institutionalized processes (Villasante 1994).

After ten years of experience, the CCCV has contributed to participatory territorial management in two ways. First, articulating social struggles and opening multi-actor spaces of dialogue and proposal co-creation. CCCV conceives society as a network; hence, participatory management from both public institutions and civil society must have the ability to articulate various social actors. For this reason, CCCV, as a social broker, drove crucial involvement of the local community in decision making. This is a matter of helping people grapple with situations of power, establishing bridges between groups, policy-makers, and professionals in different disciplines (Jacobs 2014).

Before the local government elections in 2019, the CCCV carried out the initiative 'Agenda Ciudadana'. It held an open-participative, bottom-up process that called the citizen to shed light on interest topics. The methodology was focused on identifying problems, proposing needs-driven solutions, and overcoming prefabricated solutions or tendencies through open citizen forums and discussion tables (CCCV 2019)

This article shares the experience of the 'Agenda Ciudadana', in particular on cultural and heritage management. It offers a participatory management model action-oriented, dealing with real-world problems, and enhancing civil society's decision-making. In what follows it lays out the main CCCV concepts related to heritage and cultural management. The second section explains the methodological approach used in the forums held throughout the Agenda Ciudadana. The third section delineates the problems identified and the possible solutions for them. To end up with a contextualization of the heritage and cultural management during the recession and health care crisis due to the Covid-19 pandemic.

It is essential to point out that the information presented in this article is part of the collective construction of knowledge — both from the internal reflections of CCCV and open citizen participation.

---

[1] Citizen Agenda

[2] Corresponsabilidad solidaria: Joint responsibility comes from a dialogic ethic, meaning, ethics based on the relationality of human coexistence, which enables joint responsibility, being also the commitment of the person to the community, based on intersubjectivity and an awareness of the assumption of the consequences and collateral effects of human actions (Michelini, 2007, p. 39–40).

---

[3] Código Orgánico de Organización Territorial Autonomía y Descentralización (2010).

2 CCCV KEY CONCEPTS ON CULTURAL HERITAGE MANAGEMENT AND CITIZENSHIP

2.1 Heritage and culture

Cuenca is undoubtedly an important territory for cultural encounters because of its Cañari, Inca, and colonial historical trajectory, but also because of being the south economic pole of the country with a close relationship with rural populations. Cuenca has been recognized for its cultural production in writing, theater, crafts, and its popular traditions (Durán 1999). The declaration of WHS (1999) is a milestone in the life of the city, it exalts the architectural condition of the colonial "urban landscape" of the city center consisting mainly of republican and modern buildings and part of the archaeological remains of prehispanic cultures (Jamieson 2008). This denomination has influenced the urban imaginary. Above all, it has fed budgets, public space regulations, urban plans, and defined narratives and symbolic repertoires based on an idealized image of the city (Salgado Gómez 2008). Thus, there is a notion of heritage that is broadly linked to the material, monumental, and architectural. This has focused on conservation policies on the preservation of their physical components from a static logic, which has affected the urban dynamism present in the daily practices linked to the popular and community (Cabrera 2020).

Within the broad scope of culture, several comprehensions of heritage have been developed. However, some are more dominant than others – e.g. UNESCO's -, the institutionalized discourses and uses of heritage and culture from the public administrators, NGO's and tourism trends have been more visible and positioned than the popular and local comprehensions (Brumann 2018). From the global south heritage has been challenged *"as a pre-set category predicated on this pact, colonial exclusions are inevitably replicated in the process of heritagization itself"* (Herrera 2014:80). In this line, there is a growing awareness of diverging forms and experiences of heritage that questions the authorized or institutionalized notion of Heritage (Smith 2015). The critical approach to heritage brings to light the disparities between whose heritage it is, how it is produced, and what idiosyncrasy it reinforces (Hall 2004; Smith 2006). In the face of traditional heritage and cultural management, CCCV questions top-down processes of traditional cultural and heritage management, by incorporating citizen participation and multi-actor negotiation.

To position popular understanding and ways to experience culture and heritage, CCCV appeals to work from local imaginaries and discourses within the historical trajectory of the city, taking in account that imaginaries *"enact and construct peoples and places draw upon colonial and postcolonial visions of Self and Other that circulate"* (Salazar 2011:50). In that way, CCCV works under the premise of 'building overbuilt'. CCCV acknowledges the multi-layered history of Cuenca that enables it to rethink the culture and heritage. *"Once heritage is consensually defined in context, stakeholders may participate in shared visions of development that engage invariably painful histories and address ongoing processes of exclusion, as well as producing material benefits"* (Herrera 2014:80).

For this reason, CCCV conceives Heritage as a social construction, but moreover, a field of dispute. Beyond Heritage as a monument, museums, and architectural assets quite crucial in the historic center of Cuenca, CCCV points out the human dimension involved in art production and heritage safeguarding. It is dynamic and embodied in people in a multitude of ways.

"Understanding as a symbolic, subjective, processual and reflexive selection of cultural elements which are recycled, adapted, re-functionalized, revitalized, reconstructed or reinvented in a context of modernity employing mechanisms of mediation, conflict, dialogue and negotiation in which social agents participate" (Martí 2006:96).

The CCCV is committed to a more open entry on heritage and culture, which incorporates alternative forms of heritage-making, in this sense Harrison's (2015, 2018) contributions point out that *alternative heritage is the recognition of ontological plurality: that different forms of heritage practices enact different realities, and hence work to assemble different futures, ways in which each domain of heritage relates to a particular mode of existence* (Harrison 2015:24). In this frame, CCCV seeks to entangle different comprehensions and claims about the heritage that come from the hybrid (García Canclini 1989; Herrera 2014) and pot-colonial (Kaltmeier & Rufer 2016) nature of Latin American cultural landscape but also the specific socio-economic gaps and cultural identifications in the city. Heritage as a notion that responds to contemporary readings of the past (I Martí 2006) (Muzaini & Minca 2018) configures not only cultural identities but also political subjects, therefore CCCV considers heritage and culture strategic in the strengthening of political culture and a public ethic to think and imagine cities from diversity and also conflict.

2.2 Right to the City: Heritage and culture as a public good and an ethical Project of the City

Critical urban studies (Dawson & Edwards 2004; Gottdiener 2019; Sassen 2000) provide a more systematic understanding of human settlements, assuming cities as communities with an organic network dynamic, where relationships between people, its environment, and the production of space are mutually constituent (Lefebvre & Nicholson-Smith 1991). Hence the urban communities entail close relationships with belonging, identity, aspirations, and many other ranges of variables (Bauman 2001). In this framework, CCCV conceives the city as a social construction assuming the challenge is to turn cities into spaces where all its inhabitants find the material and spiritual conditions for the enjoyment of a dignified life (CCCV 2011;

Toro 1991). For this, citizens must be included in an active and conscious process of thinking and building the city. Thus, CCCV presents the city as an ethical project based on the ability to cooperate with others to transform the social order of urban life through deliberative and participatory democracy, placing the persistence of poverty as a focus of public interest and social exclusion development models (CCCV 2011–2012). Therefore, CCCV embraces the City as an Ethical Project by resignifying equity, dignity, and justice without privileges or exclusions (CCCV 2011).

Nowadays, the new demographic reality means that city populations are more diverse than ever before. It is important to understand the city as a community continued by collectivities showing the numerous realities of the city. The city as a Public Good refers to the social interweave where the territorial organization and heritage are generated throughout the history of a heterogeneous community that inhabits it (CCCV 2012). In this sense, the public is what arises from the interaction of the network. Therefore, the city itself is a public good built by the active praxis of ideas.

In this frame, CCCV states that the new approaches should overcome the notion of heritage and culture based on homogeneous communities (Jacobs 2014), but rather rethink them within the comprehension that the social reality complexity is based on the plurality of ontologies (Harrison 2018). Hence, heritage holds a diversity of historical and modern contradictions (Kaltmeier & Rufer 2016).

In this frame, heritage and cultural management requires including all the voices and fostering the participation, to build more inclusive heritage and culture narratives that recognize all the histories and overcome colonial approaches and segregations. Likewise, to assume Heritage and Culture as a public good and an ethical project of the city advocates to shift values and ethos of our identity and culture.

In the same context that CCCV considers the city as a public good, heritage and culture are produced and have continued social building, historical and spatial conditions. The fact that culture and heritage are public goods does not imply that they are exempt from historical tensions in urban life. In this line, González (2015) states that *the fundamental alienation and inequality entailed by the appropriation of common values through heritage constructions calls for a politically involved approach to heritage studies. Granting ontological status to these common values in our critical accounts and working with communities helps them acknowledge the existence of these values (p. 34).* In the same line, heritage and culture as an ethical project is a call to do a critical reading of history and recognize the claims of social movements and in particular from the indigenous (Battiste 2016) and women organizations (Withers 2015). In this frame, investigators for Cuenca propose the heritage notion recognition, linked to the citizens' experience (Kennedy 2007), pointing out the need of a transformation in the attitude of ethics over the effects of the heritagization that ends up displacing the vulnerable (Cabrera 2019), and the urgent

integral view that includes possibilities for a real good living (Eljuri 2014).

The shift must be guided by negotiation and agreements about what kind of city, community, and society citizens would like to inhabit and build. Therefore the *"Right to the City is the collective right to change and reinvent the city according to citizens' wishes"* (Harvey 2012:20). In this sense, heritage and culture must be rethought and reinvented as collective rights oriented to contribute social transformation from the local needs, worries, experiences, and claims.

2.3 Citizen co-responsibility in the building of cultural heritage and its management

CCCV understands co-responsibility as the conscious and responsible exercise of participation in the decision-making process (CCCV 2011). It goes beyond social inclusion in the design and implementation of public policies. But rather, it refers to the use, care, and production of culture and heritage in daily life (CCCV 2011). Consequently, it must be taken into account from different dimensions and scales in families, neighborhoods, collectives. Co-responsibility is crucial for citizen action. CCCV has worked to spread and promote co-responsibility as part of an inherent social convenience, creating awareness of the consequences and collateral effects in human actions (Michelini 2007). That is why all decisions and activities of the people matter in safeguarding heritage, culture, and art production (CCCV 2012).

Therefore, it is necessary to develop citizenship capacities, enhancing coordination with local governments (CCCV 2011). Since Co-responsibility implies permanent feedback and information validation, communication turns to be strategic because it contributes to the transparency of the entire process of public administration and settles bridges between social claims and public authorities (Lorite 2018). Therefore, the CCCV calls for the ethical responsibility of communication to democratize the decision-making process. But it also enables a fluid interaction between actors and institutions. Thereupon, activating co-responsibility is rooted in dynamics that promote a feeling of belonging to co-create the city (CCCV 2011).The co-responsible safeguarding of heritage and culture means sense-making, value, and use the significant elements of our identity, to give continuity and be creative with the new artistic forms, to consolidate a stable inclusive network that enables participation in every scale about heritage and cultural imaginaries. Finally, these imaginaries are a result of the citizen praxis, shaping our past-present and future identity. Finally, citizen co-responsibility is framed by the need for a new ethical paradigm based on care relationships. Knowing how to care implies establishing close relationships with others, establishing a dialogue of respect, and knowing how to listen despite differences (Toro & Boff 2009). CCCV proposes that citizen co-responsibility about culture and heritage, must be guided by care relationships, seeking stories, spaces where to recognize each other, about the owned knowledge form native cultures, and also take care of the new heritage and cultural productions since ancestral knowledge survive in them but also incorporates contemporary learning. Care implies a constant transformation and meaning-making of practices, environments, and people as agents that create culture and heritage in their interaction. And on the other hand, the paradigm of care also tries to take responsibility for culture and heritage seeking communities' sustainability in their economic, social, and environmental dimensions (Harrison 2013).

3 METHODOLOGY: CITIZEN OPEN FORUMS

3.1 *Experience exchange*

Traditional social movement approaches seek to take static, detached, and explanatory snap-shots of collective action; meanwhile, Participatory Methodologies attempt to contribute to critical reflection to change social realities (Villasante 2006). By empowering communities and encouraging active involvement in public matters. Hence, these approaches challenge both governments and NGOs' paternalistic discourses and practices (Borda 2009).

Through social complexity lenses and network analysis, participatory methodologies try to unravel the relations, actors, problems, and strengths of a community (Gutiérrez & Villasante 2007). Consequently, participatory methodologies enable us to problematize everyday life issues and acknowledge *'vivencial'* awareness. Once the network dynamics and contradictions are visible, participative methodologies proceed to diagnose, negotiate, co-construct, and transform based on small-scale changes and establish bridges between citizen initiatives and public administration. It is not a linear process and it entails constant feedback.

The case of the forums organized by CCCV before the elections of the local government allowed people to have a better idea of the city, they wanted to live in.

Participatory activity in the city is necessary, not just to create a sense of belonging and the idea of having a community inside this city, but also to create decision-making spaces. These forums were made in Cuenca, to collect needs and proposals on many scopes (CCCV 2019). Strategic issues were identified to discuss in each of the forums:

1. Mobility, connectivity, and public space
2. Water sources and environmental management.
3. Public management model
4. Social inclusion and quality of life
5. Cultural management, identity, and tourism

The forums had three moments: Introduction, where three experts gave a brief talk on the topic – each forum incorporated three specific themes. Later, the experts, together with CCCV members, facilitated the discussion on the work tables. At last, the results of the tables were shared and validated. As a result, 500 people participated in the forums, radios, and newspapers followed the process.

Specifically, in the last forum, cultural and heritage management issues were discussed, outcomes will be developed in the next section. Within participatory methodologies, it follows an outline of steps completed throughout citizen forums that are detailed below.

3.2 *Participatory mapping*

After all, Agenda Ciudadana was a networking project. CCCV works are based on active networks to achieve a greater scope in the city and to guarantee the most diversity possible. Mapping social networks identified the agents, but also, conflict, disputes, and tendencies (Nistal 2008). CCCV wanted to bring together agents from different and even contradictory perspectives. The idea was to create spaces where people can recognize the different ways to experience the city and how problems affect some groups more than others. Also, how different experiences can build up a more holistic understanding of reality and actor-based solutions.

Therefore, different groups were settled: public and private institutions; associations, collectives, and organizations; including general citizens without any affiliation. But also, the agents that could be articulating nodes with other spaces such as media, academia, environmental, and gender organizations. The success of the call resided in highlighting the importance of the citizen and its experience, making clear that CCCV was calling for a dialogue, leaving parties and ideological proselytism out.

3.3 *Collaborative diagnosis*

After presenting the topics to be discussed, the group proceeded to divide it into working tables. It was up to the participants to be part of the topics they were most interested in, making visible where the worries and interests were located. Within the first moment at the work tables, the facilitator provided questions to encourage the dialogue. Following, the participants shared their main concerns and problems from their experiences. This lays down an overview of the situation of the topics and their approaches, its main causes, and impacts. It was crucial that during this discussion, the participants interpellated each other (Montañés 2010). Therefore, facilitators did not rush, nor canceled the initiative of the participants. As a result, the collaborative diagnosis problematized the worries by acknowledging the different positions and strains of the topic. Furthermore, it settled the ground where the solutions arose.

3.4 *Co-construction*

At a second moment on the work tables, the participants tried to elucidate possible solutions, negotiate agreements, and establish commitments. The main idea was to articulate guidelines that engage citizens and different institutions and collectives equally and specify demands and solutions that must be assumed by the public administration duty.

Co-construction entails a deep negotiation that must be rooted in respect, empathy, and solidarity. The co-construction outcomes hold emotional involvement and also materializes collective rights that go beyond our differences, focusing on common points. Citizens are all interconnected, thus they are all responsible for the solution and consequently of co-constructing a reliable network of coexistence.

3.5 *Knowledge and outcomes returns*

In the last intervention, the facilitators shared each worktable's results, enabling final feedback and validation with all the participants. The information devolutions in participatory methodologies are an ethical commitment to make the results of participatory processes accessible for those directly involved and the community. Agenda Ciudadana has been widely distributed being available to everyone. CCCV delivered the document publicly to the authorities and the media and social-networks became crucial tools to democratize this knowledge (CCCV 2019).

4 RESULTS OF THE HERITAGE AND MANAGEMENT FORUMS

This section systematizes the discussion results held in the cultural management, heritage, and identity citizen forum. The discussion focused on identifying the problems and establishing possible agreements and proposals around the cultural and heritage management in Cuenca.

Heritage and art production cannot be understood detached from their living context, hence before addressing the results, follows a brief contextualization about the cultural and heritage sector in Cuenca, considering safeguarding and promotion of the tangible and intangible heritage, as well as artistic production. Cultural management means incorporating policies that answer to the symbolic and cultural dynamics in the cities, which is continually changing.

As well, it is important to locate things like popular festivities, crafts, architecture, etc. In Cuenca, the cultural and artistic production is growing, and it has got to a constant creation of theatre plays, music concerts, gallery exhibitions, etc.

The citizens' perspective survey showed that heritage management was necessary for the people. The inhabitant of Cuenca identified as a prior heritage of the city heritage the traditional crafts, the historic center, and the four rivers, in that order of importance (CCCV 2016). This shows that the understanding of the heritage of Cuenca based on citizen perception incorporates the built space, nature, and its traditions. As seen in the survey results (CCCV 2016), the cultural heritage of the city value the colonial architecture as part of this built idea of heritage. The complexity

of cultural development has put the art and cultural workers to assign efforts to a constant policy building process. The public administration models focus on heritage based on the buildings that have displaced the people even though the population points about the value of practices and memories that signify the tangible heritage (CCCV 2019).

Felipe García (2019) says that a city is a convergence of many things, the encounter of streets, people, and location, but it is not only spatial but temporal. It is also alive, the city is lived at the moment by the people that make it, which means they would encounter this deconstructed past while they inhabit the spaces (Garcia 2019). The World Heritage Site declaration has affected the way people see the city, making them understand its heritage mainly in its tangible aspect and regarding mostly its architecture. It is like building a museum in the center of the city, where cultural activities get banned (Garcia 2017).

The results of the researches held in Cuenca on heritage policies (Cabrera 2019a) show an overvaluation of what is built which does not give way to an understanding of heritage from a dynamic, changing, and inclusive notion of cultural and heritage management. It is related to tourism development, to the detriment of other daily values linked to the popular and less picturesque look of the city. In fact, in the last two decades, Cuenca's Municipality carried out a series of projects that sought the construction of an ideal image that publicizes the image of the heritage attraction and will position the city in various rankings (Cabrera 2019b). This has generated displacement and gentrification processes and the construction of scenographic urbanism (Cabrera 2019a).

This is why, in the need of creating culture and making it part of the heritage construction, participatory processes, like the one started by Francisco Aguirre in 2011. This process specifically tried to claim the public space as a public good, meaning that the local policies banning cultural expressions in sidewalks, squares, or parks should be replaced. The proposal was that thirteen spaces should be liberated for the use of the arts. Even though the document was discussed and written, it is still complicated to use public spaces for the use of the arts (Tello 2019).

It is important to emphasize the different situations that affect heritage and cultural management in a wider sense. Active participation means the understanding of a city within many different lenses, no citizen will have the same perception as another, this poses a problem where the city has to respond to all these different views of it. Getting proposals for problems located by every actor in the table.

The citizens' forums, as said before, talked about many different topics, one of them was heritage and Culture. In these forums, people representing civil organizations, public institutions, academia, etc. talked about the ideas they had for better heritage conservation and cultures and art production. They identified problems and got agreements and proposed agreements. This will be listed below (CCCV 2019).

4.1 Problems identified

"We dream about a city that appreciates its histori-cal memory, lives it's tangible and intangible heritage, takes care, enjoys and builds it constantly, a city that takes care of its inhabitants and visitors and tells a story to them" (CCCV 2019:21).

The problems identified around heritage and its management were related to the cultural policies, which the participants said were restrictive towards the use of historic zones from the city population. The group also said that the conservation of these spaces should be considered around the active participation of people, which would let these policies to be more inclu-sive and respond to real social needs. (CCCV 2019). This again was related to the need for using spaces as public goods, and not only the thought of them as a museum exhibition (Garcia 2017).

According to the results of the forum, the restric-tions had no strong justification and made the conser-vation of the tangible goods of heritage complicated. For example, in building restoration, it isn't possible to use modern techniques or materials which would pre-serve the buildings in better ways without changing its essence. The investment for heritage protection is used only in the maintenance and recuperation of the build-ings, leaving aside many other aspects of the heritage, and even forgetting the administration of these edifica-tions. According to the conversation held this makes it more difficult to understand heritage in a broader sense than the colonial architecture, having in mind the people, the intangible heritage, and the inheritance of it to the actual date (CCCV 2019).

The work tables identified that in Cuenca citizens don't have incentives to live in the heritage zones, espe-cially in the colonial center, which means the abandon-ment of houses, and the gentrification of these spaces, where the previous owners can't manage a building that needs constant and expensive maintenance, so they seem displaced (Cabrera 2019a; CCCV 2019).

Taking into account that the cultural management in the city is part of its heritage, participants found other problems around the arts and its policies. It was identified as a difficulty to create agreements within the artistic sectors (CCCV 2019). They referred to the market dynamics, saying they do not ensure an equi-table distribution of prices and contracts. Participants said the formalities, procedures, and bureaucracy asked by the local government make the manage-ment of any cultural activity complicated (CCCV 2019).

The need to understand the reality of cultural man-agers was another topic on the table. *Participants said the formalities and procedures of the local govern-ment make the management of any cultural activity complicated* (CCCV 2019:22). In this section, debts from public organizations to the artists were identified, which went from three months to three years of bills not being paid. The table concluded that this means that the local government is not only not encouraging the arts, but having debts with the artists (CCCV 2019).

Difficult to move events to rural spaces and urban peripheries, without considering the importance of rurality and its cultural development (CCCV 2019:21), another problem identified between the par-ticipants. This also means forgetting the right to enjoy and develop a culture in remote spaces.

As said before the forums process meant that the participants would find proposals for these issues which were then used as compromises from the candi-dates for mayor in 2019. In this participative process, people identified actions to make these problems bet-ter, actions that the local government would be able to take.

4.2 Agreements and proposals

After the problem identification, each table made up by members of many different social contexts and back-grounds, proposed possible action lines that could be assumed by the new elected local government some of these were:

- *The conservation of tangible heritage must be in constant dialogue with the economic and tourism sectors, so their dynamics would contribute to its enjoyment and conservation, having in mind that is only this protection of intangible heritage that could justify the existence of tangible heritage* (CCCV 2019:23).
- The conservation of the heritage spaces, both urban and rural, must be considered from the citizen appropriation of them, which would allow coex-istence, knowledge transmission, and local and economic development. Heritage conservation must work alongside the habitability promotion and qual-ity of life. Heritage spaces: buildings and historic centers must be over any other concept: life spaces. (CCCV 2019)
- Work on the education, diffusion, documentation, and training in the recovery and appreciation of the city's historical memory. The value that tangible her-itage keeps is the symbolic value of telling stories, conveying knowledge, informing of conditions of who made them, and used them before. These assets make the city. (CCCV 2019)
- The local government must be a facilitator of cul-tural ventures. This would mean improving the process for permanent cultural spaces and projects, ensuring the art democratization, and its enjoyment for everyone in the city. (CCCV 2019)
- *Create, implement, and strengthen mechanisms for the participation of the cultural sector in decision-making processes around topics of its interest, and participatory organizational spaces for the art workers* (CCCV 2019:23). This would need the creation of a cultural bureau where the politics, planning, and prioritization of actions for the sector would be defined.
- The participation of citizens in these spaces is as important as the decisions taken by the government. It is necessary to think of every city as it is. Culture is alive and is creating tools for heritage management

and conservation all the time, it is important to understand the need for these living spaces (CCCV 2019).

This process ended up in an event where the elected government signed and compromised to these proposals, and their use in their planning. There have been few actions concerning these proposals. This has meant fewer securities for the cultural sector or the heritage protection, meaning, between many other issues, that around the Covid-19 emergency arts and cultural workers had no income.

5 CONCLUSIONS

The identity and culture produced in a city are part of its heritage, making cities unique places. Understanding the city as an ethical project, as explained before, can create spaces for culture is the base of what is produced in and for the city. Constantly thinking of its people and their needs makes participatory processes crucial. This would also let society ask questions like: How are we contributing to the construction of this cultural, identity, and urban heritage? It is the citizens who should decide the topics on the city debate and build from the bases the proposals that will answer to the reality of them who inhabit the city.

Given the enormous drop in oil prices, the current recession and the Covid-19 pandemic had a direct impact on the funding for cultural and heritage management and public universities. Consequently, heritage, cultural and artistic research and production have been limited.

The pandemic, but in the long term, the recession and climate change show the limits of an economic, social, and cultural development model. The current health crisis and the lockdown imply a turning point to reflect critically on the current state of affairs (Salazar 2020), but moreover, highlighting learnings for future challenges. Communities are adapting and drawing on their living heritage to respond to the crisis. In this context, cultural heritage is a source of ancient and modern knowledge that enables us to imagine different society models, different relationships with nature, and between people, providing a source of resilience (Harrison 2015). Living heritage, as the values and practices of the indigenous Andean cultures rooted in solidarity, reciprocity, and community encouraged initiatives that respond to the urgent needs in the face of the crisis. But also support a new life paradigm questioning the structural limits of the global market and the government's models.

CCCV comprehends heritage and culture as a social contribution, hence it assumes them as a common good and therefore as a collective-ethical project, rather than the particular interest of elites. In this sense, heritage and culture must be rethought and rephrased, also as collective right oriented to contribute social transformation from the local needs, worries, experiences, and claims.

It is important to note that the underlying premise of participatory management is to keep open-mindedness and empathic with others. Thus, to overcome singular and authoritative discourses, by a political culture based on the citizen co-responsibility. As a consequence, cultural and heritage management turns to be a political praxis from the community's engagement, guaranteeing the continuity of heritage in its meaningful sense (De la Vega 2020).

In Cuenca, the WHS declaration has settled a stattige and traditional safeguarding model of heritage mainly focused on architecture, leaving behind the importance of intangible heritage as local festivities, handcrafts, and other heritage expressions.

The open citizen forums as a participative process problematize everyday life issues. As a result, it incorporates demands and proposals from a social-basis and the popular comprehensions, knowing that heritage derives from the art and cultural identity of a city.

Agenda Ciudadana is a great contribution to strategic planning processes. It provides a holistic approach to the most important problems and demands of the population. It can be used by public administration but also by other institutions and organizations. Thus, it can be a guideline for Cuenca.

A broader analysis of the current situation is provided, including the Covid-19 pandemic, and how it affects the heritage safeguarding and the cultural management in the city. Understanding that the weakness of policies and safeguards have created a crisis in the sector, made the art and culture production stop, and the safeguarding of heritage weaken.

The uncertainty of the crisis has also created the need of setting positionalities, a necessary first step when addressing problems on any scale. It is also necessary to contribute from transdisciplinary decision sectors, and cross boundaries in this process. Cultural and artistic sectors are going to be the last ones to get back to work. And it is necessarily going to be different, because of the constraints of the moment, so it would be necessary to think of the cultural, arts, and heritage management in different ways than the ones considered until now.

Understanding the city as a community of communities, it has diverse ideologies and needs, and participatory processes are supposed to understand this convergence. Public management and actions should be strongly related to the citizens' agreements, achieving this way a heritage for the future. Innovation and creativity come from the overlapping of different domains, allowing a convergence of references, paradigms, and values.

The richness of this participative experience resides in bringing together diverse actors and strengthening social-networks. It has left more issues to question and to continue working on from citizen participation. For example: How the active and participative management of culture is a public and political exercise? How could cultural and heritage management contribute to overcoming the socioeconomic, ethnic, and gender gaps in our city? How to balance between maintaining our heritage and letting the alive culture coexist? Thinking about the approach of a non-conservative heritage, not essentialist, where heritage's life and use are for everyone and it's maintained at the same time. Cuenca is a particular case, as would be any other city in the world, it can't be planned in generic terms.

REFERENCES

Asmal, D. P. 2016. Modelo de gestión del patrimonio cultural edificado basado en la participación ciudadana para la ciudad de Cuenca.

Battiste, M. 2016. Living Treaties. Cape Breton University Press.

Bauman, Z. 2001. Community: Seeking Safety in an Insecure World. Cambridge: Polity Press.

Bolay, J., & Rabinovich, A. 2004. Ciudades intermedias:¿ una nueva oportunidad para un desarrollo regional coherente en América Latina?. DILLA, H.(coord.).

Borda, O. F. 2009. La investigación acción en convergencias disciplinarias. Revista paca, (1): 7–21.

Borja, J., & Castells, M. 2013. Local and global: The management of cities in the information age. Routledge.

Braga, M. D. S. S., & Casalecchi, G. A. 2019. Legitimidad y compromiso democrático. Impases contemporáneos en América Latina. Anuario Latinoamericano–Ciencias Políticas y Relaciones Internacionales, 7.

Brumann, C. 2018. Anthropological Utopia, Closet Eurocentrism, and Culture Chaos in the UNESCO World Heritage Arena. Anthropological Quarterly, 91(4), 1203–1233. DOI:10.1353/anq.2018.0063

Burgos Guevara, H. 2003. La Identidad del Pueblo Cañari: De-construcción de una nación étnica. Quito: Abya Yala.

Cabrera, N. 2020. El Centro Histórico de Cuenca: conservación y turismo frente a las dinámicas populares. Universidad Verdad, (76), 8–21.

Cabrera, N. 2019a. Gentrificación en áreas patrimoniales latinoamericanas: cuestionamiento ético desde el caso de Cuenca, Ecuador. urbe. Revista Brasileira de Gestão Urbana, 11.

Cabrera, Nb. 2019b. Mercado inmobiliario y metamorfosis urbana en ciudades intermedias. Gringolandia en Cuenca: la tierra prometida. Bitácora Urbano Territorial, 29(1), 91–100.

CEPAL, N. 2016. Panorama Social de América Latina 2015. Cepal.

Colectivo Cuenca Ciudad para Vivir (CCCV). 2011. Segunda Encuesta de Percepción sobre la Calidad de Vida en Cuenca-Ecuador. Cuenca.

Colectivo Cuenca Ciudad para Vivir (CCCV). 2012. Segunda Encuesta de Percepción sobre la Calidad de Vida en Cuenca-Ecuador. Cuenca.

Colectivo Cuenca Ciudad para Vivir (CCCV). 2015. Cuarta Encuesta de Percepción sobre la Calidad de Vida en Cuenca-Ecuador. Cuenca.

Colectivo Cuenca Ciudad para Vivir (CCCV). 2016. Cuarta Encuesta de Percepción sobre la Calidad de Vida en Cuenca-Ecuador. Cuenca.

Colectivo Cuenca Ciudad para Vivir (CCCV). 2019. Agenda Ciudadana. Cuenca, Ecuador.

COOTAD, C. 2010. Código Orgánico de Organización Territorial. Autonomía y Descentralización.

Cortina, A. 2008. Ética pública desde una perspectiva dialógica= Public ethic from a dialogical perspective.

Dawson, A., & Edwards, B. H. 2004. Introduction: Global cities of the South. Social Text, 22(4), 1–7.

De la Vega. P. 2020. Ecuador. Políticas Culturales y COVID-19: El desvelamiento de una crisis. RGC Ediciones. Retrived from http://rgcediciones.com.ar/ ecuador-politicas-culturales-y-covid-19- el-desvelami ento-de-una-crisis/? fbclid=IwAR0n0tDTfJlITMrkNPm 48wznB8p4VWIZhsHRu9ji1yTlYLBr9ldWReNBrE8

Durán, N. A. 1999. Artesanías, C. C. I., & Populares, A. Cultura Popular en el Ecuador.

Eljuri, G. 2014. El patrimonio cultural en el nuevo milenio. Universidad Verdad, (64), 43–68.

Garcia, F. 2017. La ciudad y el proceso de museificación. Transitando la reconfiguración territorial de la cultura. Contextos: Estudios De Humanidades Y Ciencias Sociales, (38): 145–156. Retrived from: http://revistas. umce.cl/index.php/contextos/article/view/1340

García, F. 2019. Mapping cognitive capitalism to mitigate urban museification. Perú: Revista Apuntes Editorial Pacífico. (85): 173–197

García Canclini, N. 1989. Culturas híbridas: Estrategias para entrar y salir de la modernidad. México: Grijalbo. ISBN 968-4199546.

González, P. A. 2015. Conceptualizing cultural heritage as a common. In Identity and heritage (pp. 27–35). Springer, Cham.

Gottdiener, M. 2019. New urban sociology. The Wiley Blackwell Encyclopedia of Urban and Regional Studies, 1–5.

Gutiérrez, P. M., & Villasante, T. R. 2007. Redes y conjuntos de acción: para aplicaciones estratégicas en los tiempos de complejidad social. Política y sociedad, 44(1): 125–140.

Guzmán, P. C., Roders, A. P., & Colenbrander, B. J. F. 2017. Measuring links between cultural heritage management and sustainable urban development: An overview of global monitoring tools. Cities, 60, 192–201.

Hall, Stuart. 2004. Whose heritage? Un-settling 'the heritage', re-imagining the post-nation. En The Politics of Heritage. Routledge: 37–47.

Harrison, R. 2013. Forgetting to remember, remembering to forget: late modern heritage practices, sustainability, and the 'crisis' of accumulation of the past. International Journal of Heritage Studies, 19(6), 579–595.

Harrison, R. 2015. Beyond "natural" and "cultural" heritage: toward an ontological politics of heritage in the age of Anthropocene. Heritage & Society, 8(1), 24–42.

Harrison, R. 2018. On Heritage Ontologies: Rethinking the Material Worlds of Heritage. Anthropological Quarterly, 91(4), 1365–1383.

Harvey, D. 2008. The right to the city. The City Reader, 6(1): 23–40.

Harvey, D. 2012. Ciudades rebeldes: Del derecho a la ciudad a la revolución urbana. Salamanca: Ediciones Akal.

Hermida, M., Hermida, C., Cabrera, N., & Calle, C. 2015. La densidad urbana como variable de análisis de la ciudad: El caso de Cuenca, Ecuador. EURE (Santiago), 41(124), 25–44.

Herrera, A. 2014. Commodifying the indigenous in the name of development: the hybridity of heritage in the twenty-first-century Andes. public archaeology, 13(1–3), 71–84.

Hosagrahar, J. 2018. Inclusive social development and World Heritage in urban areas1. World Heritage and Sustainable Development: New Directions in World Heritage Management.

I Martí, G. M. H. 2006. The deterritorialization of cultural heritage in a globalized modernity. notes & comments.

Idrovo, J. 1985. Tomebamba: primera fase de conquista incásica en los Andes septentrionales. Los Cañaris y la conquista incásica del Austro Ecuatoriano. La frontera del estado inca: 71–84.

INEC, D. 2017. INEC Estadísticas Económicas.

Irazábal, C. 2017. City making and urban governance in the Americas: Curitiba and Portland. Routledge.

Jacobs, M. 2014. Cultural Brokerage, Addressing Boundaries, and the New Paradigm of Safeguarding Intangible Cultural Heritage. Folklore Studies, Transdisciplinary Perspectives, and UNESCO. Volkskunde, 115: 265–291.

Jamieson, R. W. 2008. The market for meat in colonial Cuenca: a seventeenth-century urban faunal assemblage from the southern highlands of Ecuador. Historical Archaeology, 42(4): 21–37.

Kaltmeier, O., & Rufer, M. 2016. Entangled heritages: postcolonial perspectives on the uses of the past in Latin America. Routledge.

Kennedy, T. Alexandra. 2007. Apropiación y re simbolización del patrimonio en el Ecuador. Historia, arquitectura y comunidad. El caso de Cuenca. Procesos revista ecuatoriana de historia, (25). 129–151.

Lefebvre, H. 1991. Critique of everyday life: Foundations for a sociology of the everyday (Vol. 2). Verso.

Lefebvre, H., & Nicholson-Smith, D. 1991. The production of space (Vol. 142). Blackwell: Oxford.

López Caballero, P. 2008. 'Which heritage for which heirs? The pre-Columbian past and the colonial legacy in the national history of Mexico.' Social Anthropology, 16 (3):329–45

Lorite, F. M. 2018. Public cultural Planning. Periférica Internacional. Revista para el análisis de la cultura y el territorio, (19): 123–129.

Mainwaring, S., & Bejarano, A. M. (Eds.).2008. La crisis de la representación democrática en los países andinos. Editorial Norma.

Michelini, D. J. 2007. Bien común y ética pública. Alcances y límites del concepto tradicional de bien común. Tópicos, (15), 31–54.

Molina, B. 2019. Foros híbridos, participación y gestión sostenible del Patrimonio Mundial. El caso de Santa Ana de Cuenca. methaodos. Revista de ciencias sociales, 7(2), 225–243.

Montañés, M. 2010. El grupo de discusión. Cuadernos CIMAS, Observatorio Internacional de ciudadanía y medio ambiente sostenible: 1–29.

Muzaini, H., & Minca, C. 2018. Rethinking heritage, but 'from below'. In After heritage. Edward Elgar Publishing.

Nistal, T. A. 2008. IAP, Mapas y Redes Sociales: desde la investigación a la intervención social. Revista de Trabajo Social PORTULARIA: 1–32

ONU-HABITAT. 2010. Estado de las ciudades de América Latina y el Caribe. Rio de Janeiro. Programa de las Naciones Unidas para los Asentamientos Humanos.

Pérez, J., Tenze, A. 2018. La participación ciudadana en la Gestión del Patrimonio Urbano de la ciudad de Cuenca (Ecuador). Estoa. Revista de la Facultad de Arquitectura y Urbanismo de la Universidad de Cuenca, (7): 229–254.

Salazar, N. B. 2011. Imagineering cultural heritage for local-to-global audiences. The Heritage Theatre, (January 2011): 49–72.

Salazar, N. B. 2020. Anthropology and anthropologists in times of crisis. Social Anthropology/Anthropologie Sociale. European Association of Social Anthropologists. DOI:10.1111/1469-8676.12889

Salgado Gómez, M. 2008. The Cultural Patrimony as a totalizing and technical narrative of governance. OLACCHI (1): 13–25.

Sassen, S. 2000. New frontiers facing urban sociology at the Millennium. The British journal of sociology, 51(1), 143–159.

Smith, L. 2006. Uses of heritage. London and New York: Routledge.

Smith, L. 2015. Intangible Heritage: A challenge to the authorized heritage discourse? Revista d'etnologia de Catalunya, (40), 133–142.

Taylor, Ch. 1997. Las fuentes del Yo, Barcelona, Paidós.

Tello Tapia, T. 2019. La criminalización del teatro en los espacios públicos del casco urbano de Cuenca, Universidad del Azuay, Cuenca.

Timothy, D. J., & Nyaupane, G. P. (Eds.). 2009. Cultural heritage and tourism in the developing world: A regional perspective. Routledge.

Toro, J. B. 1991. Tiempos de ciudadanía, tiempos de democracia. Revista Foro, (14), 100–104.

Toro, J. B., & Boff, L. 2009. Saber cuidar: El nuevo paradigma ético de la nueva civilización. Elementos conceptuales para una conversación.

Villasante, T. 1994. Sobre participación ciudadana. RTS: Revista de treball social, (133), 48–57.

Villasante, T. 2006. La socio-praxis: un acoplamiento de metodologías implicativas. Metodologías de investigación social. Introducción a los oficios: 379–405.

Villasante, T. 2011. Estilos y epistemología en las metodologías participativas. Democracia Participativa y Presupuestos Participativos: acercamiento y profundización sobre el debate actual. CEDMA – Centro de Ediciones de la Diputación de Málaga: 123–148.

Williams, G. 2002. The other side of the popular: neoliberalism and subalternity in Latin America. Duke University Press.

Withers, D. 2015. Feminism, digital culture, and the politics of transmission: Theory, practice, and cultural heritage. Rowman & Littlefield.

The Future of the Past:
Paths towards Participatory Governance for Cultural Heritage – García et al (eds)
© 2021 Taylor & Francis Group, London, ISBN 978-1-032-02129-4

Globalization and heritage at risk: Theories for its governance

L. Herrera
Facultad de Filosofía y Letras, Universidad de Cuenca, Cuenca, Ecuador

J. Amaya & A. Tenze
City Preservation Management Project, Faculty of Architecture, University of Cuenca, Cuenca, Ecuador

ABSTRACT: This paper offers a synthetic analysis of capitalism as a global system which engenders negative effects on cultural heritage. Accordingly, the first aim of the paper is to systematize a wide variety of information on heritage assets which are at risk. As a response to such jeopardizing circumstances for cultural heritage, a second goal of the paper is to suggest governance processes as valuable mechanisms for the protection, preservation and promotion of such assets. The research methodology is based on hermeneutics, an approach that enables the rigorous interpretative analysis of information regarding heritage. The text is organized in four parts: the introductory section discloses the problematic, justification and purposes of the study, and explains the conceptual approach to heritage; the second part presents the methodology that has been applied, and the third part exposes the research findings. Finally, the conclusions synthesize evidence on how capitalistic globalization clearly jeopardises cultural heritage.

1 INTRODUCTION

The globalizing logic has strengthened capitalist social relationships and its dangerous economics impositions, which nowadays social scientists refer to as neoliberalism. Certainly, other civilization processes have also imposed their own logics and hallmarks, but none of them with such broad effects on a planetary scale. In this context, the neoliberal market society called by Stiglitz (2011) as "free fall"-, and the liberal oligarchies -as pseudo democracies (Castoriadis 2002)- prevail as the only civilizational perspectives. While most of the planet's population survives in conditions of poverty, the worldwide monopolistic accumulation of wealth is currently quantified in trillions of dollars. Nowadays the world responds to a hegemonic history which contradicts the people's self-government and, hence, promotes elite's minority and totalitarian socio-political control. As a consequence, globalization has placed the planet in one of its deepest crises. As Beck (1998) declares, the contemporary globalized world has become a society at risk.

Global capitalist supremacy significantly affects the ecological balance. As confirmed in the annual Bulletin of the World Meteorological Organization (WMO), between 1990 and 2013 climate warming increased by 34% due to long-lasting greenhouse gases, such as carbon dioxide (CO_2), methane (CH_4) and nitrous oxide (N_2O) (Herrera 2016). At present there are 415 parts per million (ppm) concentration of CO_2 in the atmosphere, when the natural level of the Earth has been, for millions of years, 280 ppm.

Another serious harm to planetary ecology is directly linked to deforestation levels and water source wear. "More than half of all terrestrial biodiversity lives in forests; between 2000 and 2012 more than 2.3 million square kilometers (230 million hectares) of forests have been lost. Among the countries with the greatest losses are Russia, Brazil, Canada, USA and Indonesia" (Herrera 2016).

Water scarcity already affects all continents. Nearly 1.2 billion people, almost a fifth of the world's population, live in areas of physical water scarcity, while 500 million move toward to such a harsh situation. Another 1.6 billion, about a quarter of the world's population, face different conditions of economic water scarcity (International Decade for Action 'Water Source of Action'. 2005–2015) (Herrera 2016).

Regarding the geopolitical scenario, contemporary societies face a series of conflicts of high intensity and global impact. In 2016 there were around 33 armed conflicts in the world. Of these, according to the Alert 2017 publication, the increasing number of deaths, injured populations, displacements, sexual violence, food insecurity, destruction of infrastructure and nature, mental health problems, among the main ones, constitute facts that should ineluctably be taken into account.

The percentage of high intensity conflicts in 2016 (40% of total wars) shows a conspicuous increment of cases if compared to that of previous years (31% in 2015, 33% in 2014). […] Accordingly, the social contexts where these conflicts took place (46%) registered, during 2016, a critical worsening of its social standards, with higher levels of violence and instability. […] Likewise, the increase of forced displacement has been more than 50% over a period of five years (as reported by UNHCR in 2016, there were 65.3 million displaced at the end of 2015, including 21.3 million refugees) (Navarro et al. 2017: 15).

As explained by Ives Lacoste (2009), these high intensity scenarios are rooted in past conflicts such as the First and Second World Wars, and the Cold War. It is from these former XX century events, and due to the Technical Scientific Revolution that they engendered, that arises the present-day undisputed predominance of capitalism. Therefore, it is war which constitutes the foundational axis of contemporary globalization.

These continued conflict and war scenarios, together with an economic logic which compels an ever-increasing productiveness under conditions of massive privatization and extractivism, have exacerbated the levels of endangerment of natural and cultural heritage in an unprecedented way. That said, in order to fully comprehend how heritage, as promoter of life and culture, is being jeopardize in contemporary contexts of global capitalism is necessary, in the first place, an adequate theoretical approach.

The concept of heritage had, for a long time, privileged museums, collections, deposits and buildings. However, this trend has lost support by the emergence of approaches that characterize heritage not only from its objective standpoint (Herrera 2013). Heritage is nowadays defined as a cultural and social product which for the most part entails symbolic meanings (Prats 2004) and representations relevant to community's identity processes. Within these conceptual considerations, heritage pose contradictions, confrontations, hegemonies and subalternities, in short, power relations and conflicts.

Heritage, to the extent that it intends to represent a reality, constitutes an inevitable field of symbolic confrontation that can take place both between different and concurrent versions of a specific cultural production, and/or between diverse and antagonistic social groups that come into contact. Since symbolic patrimonial referents usually have material support, it is not surprising that conflicts often lead to obvious material actions such as the destruction or appropriation of heritage (Prats 2004).

All cultures build meanings in strict adherence to language. Reality, in order to be understood and explained, requires complex interrelations of senses which, only through the mediation of language, enable the interpretation of the world in a coherent way. However, it is important to point out that a single language is not enough: the abundant plurality of communities and signs are dynamically constructed in never-ending sociocultural processes. Thereupon, heritage responds to cultural and linguistic conditions, linked to the historical memory of the people and a constant resignification built under different communicative processes, which currently integrate orality, writing, and information technologies, among others, and certainly pose substantial differences with respect

to earlier times. Consequently, language intervenes directly in the symbolic configuration of the world, with a direct impact on the preservation of cultural heritage and legacies. Heritage implies survival and social legitimacy and, therefore, it is a priority component of cultural identity of communities and peoples.

Since the late 1970s, UNESCO has built a non-reductionist and operational definition of heritage (Kurin 2004; Van Zanten 2004) that aims to integrate its tangible and intangible components. Cultural heritage is not an issue that can end up breaking the subject from the object, the materiality from the immateriality, the modern from the traditional, as it actually has happened in scientific approaches of Cartesian positivist nature. Cultural processes are, conversely, a never-ending complexity which binds together existence and interpretation and surpasses Cartesian approaches.

Along these lines, a religious temple, for instance, becomes irrelevant if not connected to the rituals that are performed in it and which bestow it with social significance. Likewise, archaeological sites would not have any value if they were considered detached from its historical processes and representations in social memories. Festivities must be considered in the same manner: its social meaning is deeply connected with crafts, costumes, foods, stories, parks, and a vast number of other elements that are part of their material display. Even oral productions need narrators who follow pragmatic norms in order to convey the stories and its meanings; without them languages would not acquire any concreteness.

Regarding the sustainability of heritage, its permanence cannot be simply explained in terms of hegemonic practices of certain dominant groups within societies. Nonetheless, it must be taken into account that globalization has allowed western culture to impose a large number of its hegemonic models producing the irretrievable loss of ancestral knowledges and even languages that constituted fundamental heritage referents of numerous indigenous and Afro-descendant cultures. It is compulsory, therefore, to integrate cosmopolitan interpretations (Beck 2005) and intercultural perspectives (Fornet-Betancour 2009; Tubino 2015). In this respect, this research aims at systematizing information on assets at risk and suggesting governance process in order to protect, preserve and promote heritage.

2 METHODOLOGY

Gadamer (1977) explains hermeneutics first and foremost as interpretation -which thus necessarily responds to epistemic and subjective processes- and synthesizes it mainly as interpretation of texts. In accordance with such a methodology, this paper analyses several of UNESCO reports regarding the increasing endangerment of humanity's heritage; at this respect, it is important to note that UNESCOs database contains some of the most relevant information concerning worlds cultural heritage.

The paper's hermeneutical procedures focus its interpretative undertakings on two main components: the review of relevant newspaper information on heritage at risk, and the underpinning of these data through specific theories that deal with the topic. Since it is obviously not possible to include all and every heritage asset at risk in the world, this research addresses only those that carry outstanding meaning and significance.

Consistent with what was expounded in the introductory section, its emphasis is placed on the heritage assets that are being endangered by economic practices of extractivism and by the constant escalation of high-intensity and war scenarios. In this regard, the research includes the analysis of data dealing with contexts such as: the environmental impact of forest fires in the Amazon ecosystem; the destructive wars and its consequences on cities of ancient civilizations in nations as Turkey, Afghanistan, Iran, Iraq, Libya and Syria, and the negative effects of some tourist practices that jeopardize important world heritage sites like Machu Pichu and the Galapagos Islands.

3 RESEARCH RESULTS

The contemporary economic dynamics, linked to neoliberal paradigms and productive practices based on privatization and extractivism of natural resources and biodiversity, have endangered the heritage of a larger number of peoples, compromising some of its most important referents of identity construction many of which have survived already for thousands of years. By so doing, capitalistic globalized practices give way to transnational-business-government dynamics in areas where tourism, pharmacology, hydroelectricity, mining, hydrocarbons, among the principals, represent significant monetary profits for transnational corporations. As a result of such a voracious economic behavior, in 2016, UNESCO considered that no less than 48 world heritage assets were in serious risk (UNESCO 2016). Within these, UNESCO highlights emblematic sites such as Taj Mahal, Machu Pichu and Park Güel, which are being critically affected due to touristic activities that are conducted without any sort of regulation or control. Additionally, 152 sites were identified as vulnerable heritage assets due to environmental pollution, tourist pressure, real estate ventures and commercial impact. In Africa, for instance, pollution problems, deforestation and political conflicts (Cerodosbé 2018) have caused the destabilization of national parks such as Manovo-Gounda St Floris (Central African Republic), Garamba, Salonga and Kahuzi-Biega (Democratic Republic of the Congo).

By the same token, the terraced rice fields of the Philippine mountain ranges, declared as World Heritage by UNESCO in 1995, are currently considered endangered. These terraces, built by the indigenous Ifugao people, consist of a network of springs that channel water from subtropical forests, creating an irrigation system and an agricultural legacy emblem,

transmitted from generation to generation. Currently, many terraces remain at risk since they cannot cope with the impacts generated by an ever-growing number of visitors. Additionally, the levels of vulnerability of these assets increase due to the emigration of young Ifugao farmers to the cities (Fernández 2013).

Correspondingly, the UNESCO World Heritage Committee unanimously decided, in June 2018, to keep the old city of Jerusalem and its walls in the list of World Heritage in Danger, since the uncontrolled urban development, the unconstrained tourism activities, together with the lack of maintenance and conservation measures, were resulting in the severe state of deterioration of religious buildings and other heritage assets (EFE Agency 2018; Velasco 2012).

In the Kurdish city of Hasankeyf, Turkey, a team of archaeologists discovered the palace of a former neo-Assyrian governor, dating from the eleventh to seventh centuries B.C. In the site, mural paintings were preserved, along with an installation for an oven on wheels and tombs with treasures (Johannes Gutenberg Mainz University 2008). The city in question constitutes historical evidence of 10,000 years, in the Neolithic era (Collective for the social revolution of Rojava Kurdistan peace and freedom, 2016), and it is also a priority symbol of the history of peoples such as Hurrites, Mitanos, Assyrians, Urartians, Medes, Persians, Romans, Byzantines, Umayyads, Abbasids, Seleucids, Artúkidas and Ayubíes. This world heritage site is considered at risk due to the anti-Kurdish policy that different governments have implemented, and the construction of the new Turkish dam of Ilisu, which constitutes a serious potential cause for flooding vast territories and, consequently, for promoting future displacements of thousands of people from their livelihoods (AFN NEWS 2019; HYG 2019).

As for the American continent, in 2012, UNESCO included the Fortifications of the Caribbean Coast of Panama (Portobelo and San Lorenzo, province of Colón) in the list of World Heritage in danger, for the following reasons: environmental factors, lack of maintenance and uncontrolled urban development (UNESCO 2016). The lack of funds for its proper protection is provoking the fast deterioration of the sites. For similar reasons, in 1986, the Chan Chan Archaeological Zone (Huanchaco-Peru District) was also included by UNESCO in the list of heritage places at risk. In addition to deteriorating environmental conditions, illegal agricultural practices and uncontrolled urban growth, the situation was much more complicated by the devastation caused by the El Niño phenomenon. In 2016, the territorial landscape was aggravated by the construction of an animal feed plant, mostly affecting the territory and its heritage characteristic (UNESCO 2016).

In 2001, a warn of threat was issued for the Galapagos Islands (Ecuador), a natural heritage site of Humanity. The warn resulted from the combination of a number of factors that jeopardize the island's environment: contamination of beaches, use of pesticides in agriculture, transformation of water bodies, expansion of the agricultural frontier, deforestation of forests and global warming effects (Alarcón 2017). This threat was radicalized in 2019, by the allowance to the United States aircraft carrier to carry out operations against drug trafficking and piracy control in the zone. The consequences can be extremely severe for the islands' biodiversity: "Environmental groups, on the one hand, claimed that the expansion of the airport could have a negative impact on the species that inhabit the islands, many of them unique on the entire planet and in a weak biological balance" (BBC 2019).

The last great tragedy on American soil occurred in 2019 with the fires in the Amazon that devoured more than 7,000 square kilometers as a result of various entropic actions, such as the change of land use, the decrease in financing for the protection of the Amazon, and weak legal enforcement of surveillance (Alencar et al. 2019, Silva et al. 2020).

The former quotation invites us to discuss the issue of wars and their implications over heritage. Contemporary high intensity conflicts have unprecedent impact on peoples' histories and on the link they maintain with their territories, whether urban or rural. Even if the process of destruction ended, any means of reconstruction would be extremely expensive and, in future predictable scenarios of high vulnerability, there would not be any guarantee that crucial heritage emblems could be safeguarded.

Within this global war scenario, the present research deals exclusively with the foreseeable damaging impacts on historical and heritage buildings. The situation in Iraq, Afghanistan, Libya and Syria has already been highlighted by UNESCO (2016). In the conflicts that have taken place in all these territories, global actors, most of them external to the aforementioned regions, have played a major role. The UN, despite being the organism that represents the world's plurality of nations, has lacked the capacity to offer effective preventive and protective measures. Conversely, confrontations have been unleashed by contradictions between powerful interests of global actors in an ever-more globalized world, and the resistance of local peoples and communities to the global exercise of hegemony. However, those resistances are often reactive actions that lack integral perspectives based on the pursue of peace, cosmopolitanism and interculturality.

In Iraq, an important part of the millenary heritage has already been destroyed and virtually all its cities would have to be rebuilt; in order to reach such an aim, obviously, a major national and international investment would be required (Concepción 2018). Additionally, some of Iraq's declared World Heritage Sites -for instance, the fortified city of Hatra- have already been registered in the list of endangered heritage assets.

Afghanistan, another scene of war, has also undergone profound impact on its heritage. UNESCO highlights a religious an ensemble which "shows the artistic and religious creations characteristic of the ancient Bactriana between the 1st and 13th centuries. The place includes several monastic ensembles and Buddhist sanctuaries, as well as fortified buildings from the Islamic era" (UNESCO 2016). This specific historical and religious asset is, since 2003 is on the list of heritage in danger.

The situation Libya is rather similar: the confrontation between the forces connected to Muammar Gaddafi and those of the opposition prompted UNESCO (2016) to declare the ancient city of Sabratha at risk, as well as four other Libyan sites.

In Syria, since 2011, war has cost thousands of human lives and the destruction of heritage rooted on four centuries of influence of ancient Babylon, Egypt, Persia, Greece and Rome. During 2014, the armed conflict affected 24 sites (with damages of between 75% and 100%), some of which were considered World Heritage by UNESCO. Among them are monuments 7,000 years old; castles of the time of the Crusades; the archaeological site of Palmira, and to the ancient cities of Damascus and Aleppo. In addition, 104 sites are severely damaged (between 30% and 50%) while other 85 suffered moderate damage (between 5% and 30%).

The war has also affected other regions where its impact on heritage is undeniable. In Latin America, for instance, multiple armed conflicts in Colombia have caused a great number of indigenous communities, some of them from the Amazon rainforest, to be forcibly displaced from their ancestral territories, terrains vital not only for the social and economic survival of its peoples, but also sacred and imbued with profound cultural symbolisms.

As has been expounded, the data that have been studied on UNESCO reports and other contributions about neoliberal globalization, have demonstrated a dangerous influence of such a system over humanity heritage. Those studies showed anti-ecological productive processes, under the logic of intensive extractivism of natural resources and biodiversity, the destruction of emblematic buildings by high intensity wars, and the extinction ancient symbolisms as a result of aggressive colonialist impositions.

That said, this paper proposes heritage governance as a means of overcoming the state of unprecedent risk that heritage faces at the current time. That is, it suggests the enhancement of governance processes in order to protect, promote and preserve the world's natural and cultural heritage.

Alejandro Suárez (2002) conceives conventional and unconventional approaches to governance. The problem that Suárez points out is that, in most cases, citizens do not influence or have a role in decision-making and, therefore, in power generation. Unconventional proposals, on the other hand, consider citizen participation crucial, not only regarding government decisions, but also for the exercise of social control and accountability, on every instance of the State. As Suárez understands it, when governance is considered an exercise of power solely from the State, without the inclusion of citizens, citizenship is limited to the status of citizens as voters and not as active subjects of government.

According to Nef (1992) (Suárez 2002), governance should be understood as the horizontal relation established between rulers and governed. In this sense, governance is about understanding processes through which state and government agents are subject to "accountability" dynamics. Nef proposes, therefore, governmentality or governance, since in the usual idea of government the representation of the people is insufficient and its participation in decision-making is restricted. Participation must include accessibility, equity, justice, and values of democracy. Nef argues that governmentality refers to the endogenous ability to self-govern, because empowerment and problem solving involve self-managed citizen participation. However, he also recognizes that the demands often exceed the government's ability to meet and solve them. In that direction, he proposes a dialectical relationship between ends and meanings, which must be deeply interconnected. That is, development, in this case, is presented in terms of purposes, capacities and opportunities present in the communities themselves and not as linear and standardized procedures. The conceptual distinction between governance and governability is a current debate (Vicher 2014). Governance should seek as a final task the institutional stability that allows it to give sustainability to territorial relations that, based on the construction of consensus, allows to deliberately act and strengthen democratic processes of territorial understanding, to legitimize practices and agencies of diverse social actors.

The complexity of global power management causes systemic conflicts, which even endangered humanity's survival (Posadas 2015). Cultural heritage is not exempt from this conflictive predicament. From the perspective of Finkelstein (1995) it is clear that the concept of global governance, which emerged from a United Nations agreement, is anachronistic in its structuring and in its proceedings. The author states that such a construct should be replaced with new, more democratic approach that will result in innovative ways to reconcile differences at international level.

For its part, Saddiki (2009), proposes in the concept of cultural diplomacy a path towards the construction of a transnational cultural dialogue, supported from an approximation and structuring of networks of actors from civil society, capable of building a framework of institutional understanding at a global level, which allows to revert the conflicts of high intensity engendered today. For some sectors of the academy, cultural diplomacy is one of the key foundations of the 21st century, a foundation on which one can build a worldwide trust and mutual understanding.

One of the key pillars of planetary governance construction from below, is culture, within which many authors ascribe a preponderant place to well-being and territorial cohesion (Hawkes 2001, Nurse 2006, UNESCO 2013, Throsby 2016). This governance cannot obviate the emotional value that is generated in various territorial contexts, which are articulated in heritage assets and cultural manifestations that are a substantial part of the peoples' lives. Consequently, the social construction of the territory is based on historical heritage legacy: it is the intergenerational dialogue which maintains heritage alive while relentlessly updating it according to the innovative perspectives and practices of the new generations (Prats 2005). It is precisely from the standpoint of heritage legacies that governance dynamics, both global and local, should emerge. The activation of cosmopolitan processes and intercultural dialogues are fundamental to reverse current conflicting global hegemonies that arise from the merely extractivist economical nature of capitalism, which transform nature, landscapes and historical cities into simple generators of accumulation and wealth, no matter what they destroy in their path.

4 CONCLUSIONS

In agreement with what Urlich Beck posits about contemporary societies, it can be concluded that capitalist globalization has placed planetary social processes in an unprecedented situation of risk. The research undertaken has shown that extractive economic activities, high-intensity wars, and tourism practices that lack proper regulations and environmental control, have put a large number of heritage assets and traditions in a serious state of vulnerability. These are assets that, considering its cultural significance and outstanding value, UNESCO has incorporated into its list of humanity's world heritage.

This adverse scenario is undoubtedly linked to Western domination and hegemonic practices. The current neoliberal globalization gives way to anti-ecological productive processes, under the logic of intensive extractivism of natural resources and biodiversity, the destruction of emblematic buildings by conflicts andwars, and the extinction ancient symbolisms as a result of aggressive colonialist impositions. The claimed social supremacy of the West has in many cases produced the irretrievable loss of languages, cultural knowledges and valued artifacts which belonged to a variety of peoples around the world.

The analysis of the social contexts where the world's heritage is constantly being jeopardize, required the withdrawal from common Cartesians approaches that understand the world through dichotomic categorizations. It was especially valuable the approach to cultural heritage that Prats posits, understanding it as a symbolic legacy for the cultural identity of the peoples. From this point of view, the conceptual perspective on heritage entails an adequate articulation of its tangible and intangible components, without which any issue of social reality cannot be explained with enough rigor and sense of integrality. Along a similar logic, hermeneutics was included as the research methodology, with which it was feasible to interpret reports compiled by UNESCO, as well as information shared through newspapers.

Considering the serious situation of endangerment that cultural heritage currently faces at the global level, this paper proposes heritage governance as a means of overcoming such a negative prospect. That is, it suggests the enhancement of governance processes in order to protect, promote and preserve the world's natural and cultural heritage. Such governance dynamics should emerge from the democratic participation of civil society, both at the global level and in local territories. Thus, the safeguarding of heritage can become a citizens' responsibility that, under cosmopolitan perspectives and territorial intercultural dialogues, would integrate the legacies of diverse peoples in processes of constant intergenerational updating.

REFERENCES

Agencia EFE. 2018. UNESCO PATRIMONIO. Retrieved on July 19, 2019 from: https://www.efe.com/efe/espana/cultura/unesco-mantiene-la-ciudad-vieja-de-jerusalen-en-lista-patrimonio-peligro/10005-3662753

AFN NEWS. 2019. La nueva presa turca de Ilisu inundará pueblos y emplazamientos antiguos. Retrieved on July 29, 2019 from: https://anfespanol.com/derechos-humanos/en-2019-la-nueva-presa-turca-de-ilisu-inundara-pueblos-y-emplazamientos-antiguos-8672

Alarcón, I. 2017. Migración de las aves ponderará el cambio climático. *El Comercio*. Retrieved on July 29, 2019 from: https://www.elcomercio.com/tendencias/migracion-aves-ponderara-cambio-climatico.html

Alencar, A., Moutinho, P., Arruda, V., Balzani, C., & Ribeiro, J. 2019. Amazon Burning: Locating the Fires. *IPAM Amazonia*. Retrieved from: https://ipam.org.br/wp-content/uploads/2019/09/AmazonBurning_Locating TheFires.pdf

BBC. 2019. Islas Galápagos: la polémica en Ecuador por la autorización a aviones militares de EE.UU. a usar un aeropuerto del archipiélago en el Pacífico. Retrieved on August 2, 2019 from: https://www.bbc.com/mundo/noticias-america-latina-48668877

Beck, U. 1998. *La sociedad de riesgo. Hacia una nueva modernidad*. Buenos Aires: Editorial Paidós.

Beck, U. 2005. *La mirada cosmopolita o la guerra es paz*. Barcelona: Editorial Paidós.

Castoriadis, C. 2002. *Figuras de lo pensable. Las encrucijadas del laberinto VI*. México DF: Fondeo de Cultura Económica

Cerodosbé. 2018. La masificación amenaza a los Patrimonios de la Humanidad. UNESCO advierte que aumenta el número de sitios que padecen de un exceso de turistas y que degradan su valor patrimonial. Retrieved on July 29, 2019 from: https://www.cerodosbe.com/es/destinos/masificacion-amenaza-patrimonios-de-la-humanidad_569708_102.html

Concepción, E. 2018. Cultura milenaria vs. bombas y masacres. *Granma*. Retrieved on August 2, 2019 from: http://www.granma.cu/mundo/2018-12-26/cultura-milenaria-vs-bombas-y-masacres-26-12-2018-20-12-58

Fernández, A. 2013. Las terrazas de arroz de Banaue y Batad. Rutas por estos asombrosos paisajes en las montañas del norte de Filipinas. *El País*. Retrieved on July 29, 2019 from: https://elviajero.elpais.com/elviajero/2013/05/03/actualidad/1367578865_573539.html

Finkelstein, L. S. 1995. What is Global Governance? In *Global Governance: A Re-view of Multilateralism and International Organizations, Academic Council on the United Nations System*. Vol. (1) Num. 2. Estados Unidos: Lynne Rienner Publishers.

Fornet-Betancourt, R. 2009. Interculturalidad en procesos de subjetivación. *Reflexiones de Raul Fornet-Bentaourt*. México. D.F: Consorcio Intercultural.

Gadamer, H. 1977. *Philosophical hermeneutics*. Los Ángeles: University of California Press.

Hawkes, J. 2001. *The Fourth Pillar of Sustainability: Culture's Essential Role in Public Planning*: 1–7. Australia: Common Ground Publishing and Cultural Development Network.

Herrera, L. 2013. Patrimonio cultural inmaterial en Mera, una experiencia etnográfica. Revista Ánfora. Vol. 20, N°. 35:65–91

Herrera, L. 2016. Cultura política y crisis civilizatoria. In Francois Houtart (ed), *Cambios de las culturas. Ingeniería cultural y pedagogía*: 117–149. Bogotá: Ediciones Desde Abajo.

HYG. 2019. La presa de Ilisu destruirá la ciudad histórica de Hasankeyf. Retrieved on July 29, 2019 from: https://rojavaazadimadrid.org/la-presa-de-ilisu-destruira-la-ciudad-historica-de-hasankeyf/

Johannes Gutenberg Mainz University, 2008. Descubren un palacio asirio milenario amenazado por una presa. *SINC Agency*. Retrieved on July 29, 2019 from: https://www.agenciasinc.es/Noticias/Descubren-un-palacio-asirio-milenario-amenazado-por-una-presa

Kurin, R. 2004. Safeguarding intangible cultural heritage in the 2003 UNESCO Convention: a critical appraisal. In *Museum* (No. 221–222) Vol. 56 (No. 1–2): 66–77.

Lacoste, I. 2009. *Geopolítica. La larga historia del presente*. Madrid: Editorial Síntesis.

Navarro, I., Royo, J., Urgell, J., Urrutia, P., Villellas, A., & Villellas, M. 2017. *Alerta 2017! Informe sobre conflictos, derechos humanos y construcción de paz*. Barcelona: Escola de Cultura de Pau, UAB.

Nurse, K. 2006. Culture as the fourth pillar of sustainable development. *Small States: Economic Review and Basic Statistics*. 11: 28–40.

Posadas R. 2015. Apuntes sobre las reflexiones teóricas de Ulrich Beck. *Quaestio Iuris*. (8. 10.12957/rqi.2015.18822).

Prats, L. 2004. *Antropología y patrimonio*. Barcelona: Editorial Ariel.

Prats, Ll. 2005. Concepto y gestión del patrimonio local. *Cuadernos de Antropología Social*. N° 21. Universidad de Buenos Aires: 17–35.

Saddiki, S. 2009. El papel de la diplomacia cultural en las relaciones internacionales. *CIDOB D'Afers Internacional* (88): 107–118. Retrieved from: http://www.jstor.org/stable/40586505

Silva, C., Celentano, D., Rousseau, G., Gomes de Moura, E., van Deursen Varga, I., Martinez, C., & Martins, M. 2020. Amazon forest on the edge of collapse in the Maranhão State, Brazil. *Land Use Policy*. Vol. 97. Retrieved from: https://doi.org/10.1016/j.landusepol.2020.104806

Stiglitz, J. 2011. *Caída libre. El libre mercado y el hundimiento de la economía*. Madrid: Santillana Ediciones Generales, S.L.

Suárez, A. 2002. Gobernabilidad: algunos enfoques, aproximaciones y debates actuales. En VII Congreso Internacional del CLAD sobre la Reforma del Estado y de la Administración Pública. Lisboa. Retrieved on October 10, 2008 from: http://unpan1.un.org/intradoc/groups/public/documents/clad/clad0043431.pdf

Throsby, D. 2016. La cultura en el desarrollo sostenible. *UNESCO. Repensar las políticas culturales: Seguimiento de la Convención de 2005.*

Tubino, F. 2015. *La interculturalidad en cuestión.* Lima: PUCP – Fondo Editorial

UNESCO. 2013. La cultura: clave para el desarrollo sostenible. *Congreso internacional.* Hangzhou, China.

UNESCO. 2016. 48 Patrimonios de la Humanidad en Peligro. Retrieved on July 29, 2019 from: https://seraporlugares.wordpress.com/2016/03/28/48-patrimonios-de-la-humanidad-en-peligro/

Van Zanten. W. 2004. Constructing new terminology for intangible cultural heritage. *Museum* (No. 221–222) Vol. 56 (No. 1–2): 36–44.

Velasco, H. 2012. Las amenazas y riesgos del patrimonio mundial y del patrimonio cultural inmaterial. *Anales del Museo Nacional de Antropología XIV*: 10–19.

Vicher, D. 2014. *El laberinto de "governance". La gobernancia de los antiguos y la de los modernos.* México D.F.: Instituto de administración pública del Estado de México; A.C.

Lessons from territorial participatory management for effective participatory governance systems in the cultural heritage

The Future of the Past:
Paths towards Participatory Governance for Cultural Heritage – García et al (eds)
© 2021 Taylor & Francis Group, London, ISBN 978-1-032-02129-4

Smart specialisation strategies and governance on cultural heritage

J. Farinós

Departamento de Geografía e Instituto Interuniversitario de Desarrollo Local, Universitat de València, Valencia, Spain

ABSTRACT: This paper presents arguments and proposals on how to make cultural heritage, understood from a territorial perspective in relation to cultural landscape concept, as asset on which to base Smart Specialization Strategies. These should be adapted to each context according to their own cultural and heritage, whose most common identification are tourist destinations. It is organized in four sections. First presents arguments on how and why territorial heritage is a key vector for regional and local development. Also why it should be handled with unity of criteria in planning instruments. In the second one, it is presented as part of the available territorial capital, and as key issue for socio-territorial resilience; based on new forms of governance and innovative methods of citizen participation. It the third relations between cultural heritage and territorial intelligence, social innovation and a new strategic territorial planning are addressed. It closes with a final fourth section of synthesis of main ideas, arguments and examples presented.[1]

1 TERRITORIAL HERITAGE, A TRANSVERSAL VECTOR OF DEVELOPMENT

The possibilities of territorial heritage as energizer of resources and promoter of a new sustainable development (based on activities such as leisure, tourism and the improvement of the quality of life of citizens …) are beginning to be explored (Montufo 2017). In order to move forward in this direction (and aspire to another development model) it is necessary to understand the heritage in a unitary way, planning and managing it as complete (integrated) "Territorial Heritage System".

In view of the social valorization of the territory and development projects, the natural, cultural and landscape heritage must be managed with a single criterion unit (OSE 2009). However, in this task a series of difficulties and conflicts arise. One of them is what is considered as territorial heritage and how it is being used in planning instruments (Troitiño 2019).

When talking about heritage we refer to those elements of diverse nature to which society recognizes a relevant value; both material and intangible, as advocated by the Convention for the Safeguarding of the Intangible Cultural Heritage (UNESCO 2003).

Heritage assets are 'the soul' of each territory, through which to grant new functionalities (tourist and leisure, cultural, landscape or environmental) and to assign economic value. For example, in the form of differentiated heritage services on which to achieve higher-order competitive advantages, in order to face the progressive processes of standardization and banalization (towards a unique model of modernization, production and consumption) caused by economic globalization (that increases both social inequalities and environmental risks).

As a reaction, and in the face of the threat of losing the value of places, globalization has led to wanting to strengthen them. For a sense of resistance (resilience), but also because economic globalization needs territorial 'anchors'. They serve and support globalisation on places, according with their own and useful characteristics ('territorial capital', 'attractiveness'), that limit the hypermobility of capital (Brenner 1999; Cox 1997; Keating 1997).

Hence the coined concept of 'glocalization' (Swyngedouw 2004). Hence, also, the need for new forms of government (of territories) and management (of policies aimed at facilitating their development); in sume, of new forms of governance. Both multilevel governance (Hooghe & Marks 2001; Marks & Hooghe 2004), to which more attention has to be paid, as well as horizontal, looking for cros-sectoral coordination to improve coherence of actions. And all that with a marked participatory/deliberative approach, to recover democratic values and make possible the desired territorial governance (a true rule of law) (see Farinós 2015; Romero & Farinós 2011).

In all this context, the formulation of territorial projects from a patrimonial perspective (based on environmental and cultural values) cannot avoid the interdependencies between economy, culture and nature. The reading of cultural heritage as a source of wealth gencration has opened new possibilities to explore, but

---

[1] This paper has been prepared on the basis of author' previous works (FARINÓS, 2016, 2017a, 2017b, 2019a).

DOI 10.1201/9781003182016-10

it must be done with caution (see the Brussels Charter (2009) on the role of Heritage in the Economy).

The aforementioned 2003 UNESCO Convention for the Safeguarding of the Intangible Cultural Heritage aims to preserve it, but also aims to ensure its viability and optimize its potential for sustainable development. UNESCO provides support in this area to Member States by promoting international cooperation for safeguarding, and establishing institutional and professional frameworks favorable to the sustainable preservation of this living heritage.

This can have impact on the territory and on the cities, for example by creating a system of protected sites. In the case of cities, it is worth highlighting the debates that the Vienna Memorandum on "World Heritage and Contemporary Architecture – Ordering the Urban Urban Landscape" (2005), as well as UNESCO Recommendations on the Historic Urban Landscape (2011). In that case, as Troitiño (2019) points out, one can wonder: What role should the heritage system play in spatial and urban structures? It is important to locate the landscape, and in particular heritage landscapes of special interest (because their intrinsic values and for their social appropriation), in the axis of the spatial and urban planning.

From the point of view of the Smart Specialization Strategies (S3), special relevance acquires the intangible cultural heritage or "living heritage" (those practices, expressions, knowledge or techniques that each community transmits to its later generations). Intangible heritage provides communities with a sense of identity and continuity: it favors creativity and social welfare, contributes to the management of the natural and social environment and generates economic income. Numerous traditional or indigenous knowledge are integrated, or can be integrated, in health policies, education or management of natural resources, as well as in other products and services (in line with the 'Cultural economics' idea).

In sum, without prejudice to preventive initiatives, usual and necessary inventories and catalogs of heritage policies, it is the approach and priorities of landscapes of high cultural interest the one that best places itself in the path of "territorial development strategies in a heritage key" (Troitiño 2011). Usually, the use of cultural heritage has been focused on a sectoral and micro-scale (cultural objects or spaces) way, but not as an integrated perspective of such territory to be administered. However, heritage, material and intangible has to be managed in a unitary (integrated) way, in a "Territorial Heritage System" perspective, as basis for S3 and a new sustainable model of development where territorial intelligence and social innovation play a key role. The challenge then is to get in tune with economic interests through Strategic Spatial Planning and the definition of spatial visions, leading to more participatory decision making process, when choosing, applying and making an evaluation of the results/impacts of selected alternatives (Serrano 2017).

2 THE VALUE OF CULTURE AND HERITAGE AS FACTOR OF SOCIO-TERRITORIAL RESILIENCE

In modernity, individuals are homogenized and tend to isolate themselves, thus reducing their autonomy and giving rise to passive territoriality. Urban space ceases to be a public (or community produced) space for socialization, and starts to be produced according most powerful groups particular interests, through interventions and projects that can escape democratic control (Pinson 2011). Then it becomes banal or transit space (what Augé (1993) calls 'no-places' or spaces of anonymity), for which citizens do not feel rooted and losing their ownership.

Through the usual undifferentiated development mechanisms and models, based on standardization and homogenization, there is a risk of de-territorializing the population from their environment and, consequently, losing heritage stock on which to base competitive higher-order advantages. That negatively affects local development possibilities (Farinós 2014). Just the opposite of what looks for the 'Smart Specialization Strategies' (Farinós 2016). This Smart character does not only refer to technological elements (e.g. R&D from the point of view of 'learning regions' to compete better, or new information and communication technologies in the case of the new urban model of Smart Cities; but now also, in a more ambitious and strategic way, to social innovation (vid. Moulaert 2010) and to collective intelligence (Farinós 2017b and 2019).

However, fragmentation and de-territorialisation (or passive territorialisation) have their own limit (Castells 1997). Since the territory is a public and collective good, resistance movements are produced by an active citizenship seeking to regain local autonomy. This results in a territoriality that ceases to be latent (heritage and tradition, the culture itself) to become active (it pursues empowerment and can take different forms to claim that is own and differential, sometimes giving rise to identity movements). That open the door to new forms of governance (Farinós 2015), which would lead to local governance of cities and territories through this collective empowerment and autonomy (see Paasi, 2003; Raffestin 2011; Sami & Paasi 2013; Sen 1999).

We face new territorial challenges and trends that have already been identified and conceptualized. Even some indicators have been defined for their measurement and monitoring: climate change, gender perspective, health, universal accessibility, sustainable mobility…This new Culture will not be if it does not incorporate citizens in the task of construction, management and conservation of their environment, by informing, sensitizing and co-responsibilizing them for the safeguarding of their territory and their city.

However, there is still much to do to achieve an adequate and efficient management in order to dealing with them. In this sense, a positive factor is the establishment of new forms of governance and innovative methods of citizen participation. This inevitably leads to a more effective relationship between the processes of design and application of policies (formal and informal, tacit and explicit).

These are the new forms of action that are emerging to adapt the territorial model, cities, the economy, infrastructure and landscape, to the consequences and new demands in the current 21st Century. Changes in conceptual frameworks and lessons learned should be consolidated into shared practices.

2.1 *A proactive role for territorial heritage*

As mentioned in the preceding section, Territorial Heritage includes the different forms / types of heritage present in a given space: natural, cultural and artificial. Intangible Heritage is always expressed in a certain territorial and cultural context, contributing to the configuration of collective imaginary. And we must not forget that international organizations (such as UNESCO, the International Council on Monuments and Sites (ICOMOS) or the Council of Europe or the EU) promote proactive citizen participation in management, planning and safeguarding of this Cultural Heritage. Citizen participation is also considered basic when promoting cultural policies of the States. The democratic/participatory dimension is imbuing the processes of heritage production, and links the concepts of heritage and landscape as components of the identity of places.

Properly managed, cultural heritage represents an opportunity for the development of the territories, betting on a value that goes beyond the simple protection and conservation of tangible heritage elements in order to also value intangible aspects such as traditions, crafts or gastronomy, as well as including as part of this heritage the landscape and environmental elements, as issues for new development strategies.

Simply considered as a set of "reserve" elements (to be protected) has not yet provided sufficiently positive results (Serrano 2017). In this sense, cultural heritage is a concept close to others such as 'territorial asset', 'territorial capital' and 'attractiveness'. All of them related, in turn, with 'development', in its different meanings; among which the most common, despite its limited and biased nature, is associated with 'competitiveness' (much more than with 'productivity' and 'differentiation') and growth. Instead, heritage must be considered as a fundamental subsystem of territorial projects, in a purposeful and potential sense.

This new approach implies evidencing and recognizing territory as depository of a set of resources and assets, both material and intangible, natural and cultural, that are not only the expression of its identity but also the basis for its future development strategies; shortening in this development process the gaps of social inequality by seeking equal opportunities for population groups, especially equalizing rural and urban. That leads, in turn, to the concept of 'territoriality', which can be related to the existing resources/assets in a given space, as well as to the organizational capacity of the different groups of actors (collective intelligence) which would motivate social innovation capacity.

This idea of territorial heritage cannot be separated from the consideration of the landscape, of cultural assets (material and intangible) and of tourist activity as engines for local economies. As long as they become driving sectors with spill-over effects from other sectors, enlarging and diversifying the productive fabric that reaches broad sectors of the population (entrepreneurs, workers and users), by generating new added value without progressive decreasing pre-existing territorial capital and assets that support such development.

The World Heritage Cultural Landscapes, with the relevant scientific, technical, political, administrative and financial support, can become pilot centers for landscape management. Protecting, managing and ordering landscape is an implicit path to sustainable development, compatible with the socio-economic contribution that Cultural Heritage makes possible; since there is no doubt about the important relationship between assets declared World Heritage and its tourist attraction (Serrano 2017).

In sum, the best way to value and proactively conserve territorial heritage (natural and cultural) is through proper territorial, environmental and urban planning; through prudent and creative strategies and courses of action, to make the local heritage profitable as local development factor. This new "patrimonialization" is considered a relevant issue from an urban and territorial point of view, and also from the perspective of its tourist attraction. Territory, culture, heritage, landscape, environment and economy must be managed with unity of criteria, when profiling territorial development projects that integrate urban and rural dimensions.

3 TERRITORIAL INTELLIGENCE AND SOCIAL INNOVATION: SUSTAINABLE MANAGEMENT OF RESOURCES AND ASSETS THROUGH GOVERNANCE AND A NEW STRATEGIC PLANNING

3.1 *Territorial intelligence*

We understand territorial intelligence as a social resource, as part of the territorial 'capital' available in a particular place. It is the one that allows to establish a shared point of view following a first territorial diagnosis; a starting point from which one can negotiate and decide spatial visions and future alternatives. This intelligence contributes on the one hand to a better territorial culture among civil society (knowing the territory but also the willingness to want to get involved in its management); on the other, to facilitate decision-making (both through adequate legislation and instruments through which to design and develop territorial and urban policies). Ultimately, this intelligence allows us to better advance in the desired development model (which is more efficient, fair, sustainable and participatory) as well as to guide territorial model it adopts.

According to Soja (2011), as well as to Ritter since long time before, knowledge of the inhabited place is the basis for ownership. This premise allows to expand the meaning of territorial culture, beyond mere empirical knowledge of the living space. Then it is considered a social dynamic and citizen attitude, based on empowerment and a proactive attitude for decision making (Farinós et al. 2017). This citizen participation is then more efficient, based on arguments (for deliberation – in accordance with Habermas's 'homo democraticus' theory-) and responds to a critical and proactive thought (Aristotelian 'areté') (Farinós & Vera 2016).

Gaucherend, A. (2006) identifies three types of Territorial Intelligence (TI), among which the first two have had the longest path: strategic TI (creation of permanent infrastructure for strategic supervision: the traditional spatial vision); economic TI (development of products and economic intelligence services for economic innovation actors); TI for the administration of territorial communities (creation of a resource center to encourage the development of different 'measurable / quantifiable' spaces that can be typified for more appropriate management). When all three are combined as a coherent set of practices and knowledge (individual but mostly collective, tacit or explicit -according to Michael Polayni-, own or learned), TI becomes a lever of productivity and leadership for territories, activities and companies (Dumont 2018; Farinós 2019a).

3.2 Social innovation

Approaching economic problems from the perspective of participatory territorial governance allows stakeholders to expand their agenda of economic and non-economic problems that they can address. From an evolutionary perspective of the economy and socio-institutional innovations, Gallego & Pitxer (2019) argue economy and territory relationships are reciprocal and interdependent. Their coordination requires new governance practices. It is based on the hypothesis that both (economy and space) can be understood as interdependent areas of change and innovation (at different territorial scales and institutional temporalities) and that they are linked to the dominant values in each society.

From their analysis they conclude on the crucial importance of two elements to improve the effectiveness of public policies: 1) active role of local governments and adequate coordination between different levels of government (adequate multi-level governance); 2) long term institutional changes at different levels. Both lead to more comprehensive policies and a growing articulation between physical planning and regional economic development.

But as important as changes in routines are changes in values, as well as in local stakeholders ownership. Participatory governance processes can not only create new routines but also promote and propitiate changes in values, those that regulate relations between actors and economic and territorial trends. From a more traditional point of view (goods production and delivery) there are many interesting examples: e.g. regarding tourism assets and values exploitation for local economies, immigration as process improving human capital diversity and knowledge as a way to enrich local assets, partner- institutional networks, regional policies on research and development as key factors for local economic development, etc. Each time all of them are based on more participation and human capital involvement, the so called 'social innovation'. Others are more novel.

3.3 Planners and planning

In this new context of opportunities, planners use their professional skills to serve communities and address economic, environmental and cultural social challenges by helping local community residents to: develop ways to preserve and improve their quality of life; find methods to protect nature and build environment; identify policies to promote equity and equality; schedule facilities to improve services to disadvantaged communities; determine methods to deal more efficiently with the challenge of growth and development.

Planners need to work with a wide range of people in the community, in order to been able to go into depth in their knowledge of the diverse cultures and history of the different subgroups present in society. History offers planners an understanding of the past, which helps them anticipate how today's actions can affect the future. Social studies offer them an understanding of the government, companies and social organizations they must count on when implement their proposals.

Luca Bertolini (2009) sees planning as the realization of procedures and policies at the service of citizens, which enjoy basic rights that must be respected (ecological one, a decent life, work, life living place…). Citizens have much to say as potential concerned beneficiaries of actions; but also as active part of the same planning process. Healey (2007) proposes planning from an integral perspective (economic, environmental and, mainly, social -participated and contextual, local-). Vincent Nadin (2006) places planning at the center of land management, not only to regulate land use or property use, but to proactively coordinate all policies that have a territorial impact; these arguments also are shared by Lloyd & Rafferty (2013).

3.4 Territorial and urban governance

Decision about territorial development strategies (spatial visions), about their main elements and the territorial and sustainable model to which it will give rise, require inter-institutional coordination and cooperation between actors. Not only through the closed definition of a model of decision-making based on power distribution (granting securities), but also through pact and agreement ('fedus') that refines and makes such model not only possible but even more efficient. Coordination and cooperation are basic principles for new governance, adapted to each specific context, circumstance or initiative (Farinós 2019b).

Political will and the definition of a territorial development strategy and territorial model for the future, based on coherence and coordination between the interests of civil society and those of economic actors (adequate and successful stakeholders and shareholders' participation), presents several conditions or facilitators:

- To combine greater participation and coherence of the actions: Regarding participation, a real and effective public participation, although it may be conflictive, ends up being a differential element of the territorial culture that conditions (and may even limit) the new desirable practices of governance and spatial planning. Regarding coherence, it is one of the great principles of territorial governance. Despite being unquestionable, it usually has little application and practical realization. Smart Comprehensive Spatial Planning could help for that.
- Seek greater legal certainty from the new forms of governance: Governance could be understood as 'structure' (system of institutions and procedures; the framework) but also, in a more strategic way, as 'process' (the governance actions themselves). The results, the effects of such governance and its value (good governance or not, good results for the intended purpose or not) come from their combination.
- To combine, (in an intelligent and creative way, adapted to each circumstance) the traditional methods and techniques of binding/hard planning (Master Plans) with a more strategic and participated planning and management. And thus to be able to guarantee effective governance, adapted to the particularities of each territory or place. From the governance perspective comprehensive approaches to planning and management are demanded, that were selective (with top-down and bottom-up approaches). In this new governance, attitudes and values become increasingly important. A new reconstituted spatial planning, as a standardized support instrument for decision-making processes in any government action, giving rise to what we called 'full territorial governance' (Farinós & González 2019).

4 BY WAY OF RECAPITULATION AND SYNTHESIS

Planning must present integrative projects in which local aspects such as culture are respected, and must adapt to all contexts (which are also changing). For this, it is necessary a goof knowledge of places and not to impose general recipes in a de-contextualized way. Plans and projects must be consistent with the needs of people. As Gallego & Pitxer (2018) say:

"...the territory will favor the establishment of necessary cooperation to develop any collective projects that share and exchange different resources...in which how much the greater the participation and stakeholders involvement, the greater the capacity to mobilize local resources and, consequently, the success and scope of the projects". For this reason, local population participation is needed; a formula for people to regain confidence in proposed policies, plans and projects.

A territorial governance project proposed at this scale serves the locality: it is sensitive and responsible with the use of the territory, seeking cohesion, development and social integration based on their own singularities and common elements. But it also understands the global context, raises objectives that have been widely agreed and translates them into concrete programs and achievable projects, paying special attention to the potential and limitations of each territory.

This greater commitment to the territory and to the local scale (greater competencies/powers, better mechanisms and means to carry them out and improve the effectiveness of policies and actions) is the direction that political powers have followed in recent years in developed countries.

In order to achieve this, the existence of adequate territorial information is needed (free, well contrasted and truthful) on which to generate a shared understanding of real needs and possibilities of each territory (Farinós 2009; Farinós 2011). However, data and statistics, databases, new information and communication technologies from which to access, treat and disseminate them, are not enough to promote the development of territorial intelligence (despite its undeniable and essential utility to take decisions evidence based) (Flores et al. 2018).

A certain level of social innovation and new relational dynamics is essential. As Innerarity (2010) points out, it produces changes in the structures, customs, rules and processes of a given community and its territory. Something that lead us, again, to the question of territorial governance as a way of channeling the processes of collective intelligence. This capacity for agreement and commitment, promotes a new resilience (environmental, economic but also social – vid. Zapouzaki 2018-) based on a sustainable social innovation. Agreement and contractualisation would give rise to adequate territorial intelligence based in new forms of territorial governance; also, tourism governance (see Farinós 2019a; Pulido-Fernández & Pulido-Fernández 2019).

Culture and heritage (both material and intangible) have been incorporated into the tourism offer, as territorial resource for 'attractiveness' and competitiveness of a tourist destination. They are helpful to attract specific sectors of demand (tourists and trippers) who are looking for new practices and experiences. In some cases, they become part, to a greater or lesser extent, of mass tourism circuits.

Territorial heritage, besides being an asset for development, also represents an opportunity to correct inadequate socio-economic dynamics with serious negative environmental impacts, allowing territories to obtain the benefits of tourism but minimizing its impact. Beyond scientific will and social commitment, in order to configure sustainable heritage destinations, effective cooperation mechanisms between sectors and stakeholders, as well as active and responsible management of existing heritage, are needed (Troitiño & Troitiño 2018). We talk about a new tourist responsibility. Not only of a social and corporate type but, in a new and more potential way, of a territorial type, with the available resources that the touristic sector uses as a lever of development.

This could find a concretization and take the form of an ethical code for all the actors concerned, in order to become a differential fact and competitive advantage: a new model of sustainable tourism, as part of a new development model, based on territorial intelligence.

Thus, a new and adequate governance is based on territorial intelligence, a new way to understand territorial, productive and social processes. It is a new and more effective way to integrate different territorial stakeholders and their perspectives, visions and initiatives in the definition of strategies, alternatives selection and decision-making. In this sense, the convenience of combining the different types of TI (economic and territorial) as one of the best ways to open doors to a new smarter and more sustainable development model (Smart Specialization, Smart Growth) has been raised. That leads to innovation and supply differentiation (of heritage, landscape, tourism …) based on high-order advantages.

From the point of view of the necessary inter-institutional coordination, initiatives such as the Spanish National Cultural Landscape Plan (see Mata 2017) has been entrusted with a coordination task in order participation of any public administration, private entity or civil society occurs; consistent with the best conservation of the values of landscapes of high cultural interest. For this purpose, the Commission for Monitoring the Plan was constituted, in which technicians from regional governments, national government and members of civil society participate. This Commission is carrying out the very necessary task of facilitating the knowledge and dissemination of the initiatives that are being implemented by the different administrations, the shared learning of good practices, by stimulating debate and cooperation to define priorities around a collective and complex good as territorial heritage (in this case cultural landscape).

REFERENCES

Allen, J; Massey, B. & Cochrane, A. 1998. Rethinking the region. London: Routledge.

Augé, M. 1993. Los "no lugares" espacios del anonimato. Una antropología de la sobremodernidad. Barcelona: Gedisa.

Bertolini, L. 2009. The dream of planning. Planning Theory & Practice 10(3), 309–313.

Brenner, N. 1999. Globalisation as reterritorialisation: the rescaling of urban governance in the European Union. Urban Studies 36(3), 431–451.

Castells, M. 1997. The power of Identity. Blackwell: Oxford.

Cox, K. 1997. Spaces of globalization: reasserting the power of the local. New York: Guilford.

Dematteis, G. & Governa, F. 2005. Territorio y territorialidad en el desarrollo local. La contribución del modelo SLOT. Boletín de la Asociación de Geógrafos Españoles, 39, 31–58.

Dumont, G.-F. 2018. Notre vision de l'Intelligence Économique et Territoriale. La lettre d'Intelligence Économique et Territoriale de l'EM Normandie – N° 28. 'Comprendre et Emprendre'.

Farinós, J. 2009. Bases, métodos e instrumentos para el desarrollo y la cohesión territoriales. Diagnóstico y propuestas para el debate y la acción. In Farinós, J.; Romero, J. & Salom, J. (coords.) Cohesión e inteligencia territorial. Dinámicas y procesos para una mejor planificación en la toma de decisiones. Valencia: IIDL/PUV, Colección Desarrollo Territorial 7, 17–62.

Farinós, J. 2011. Inteligencia Territorial para la planificación y la gobernanza democráticas: los observatorios de los territorios. Proyección V (11), 45–69.

Farinós, J. 2014. Re-Territorializating Local Development in EU; Local-Based against Globalisation Impacts. In Salom, J. y J. Farinós (eds.) Identity and Territorial Character; Reinterpreting Local-Spatial Development. Valencia: IIDL/PUV, Colección 'Desarrollo Territorial' 13, 13–35.

Farinós, J. 2015. Aménagement et gouvernabilité. Les liens entre rhétorique et pratiques. Une dernière chance pour le projet européen?. L'Information géographique 79(1), 23–44.

Farinós, J. 2016. Planificación territorial y desarrollo local, y su relación con las nuevas formas de gobernanza territorial asociadas. Un renovado espacio de aplicación profesional. In Noguera, J. (ed.) La visión territorial y sostenible del desarrollo local. Una perspectiva multidisciplinar. Valencia: PUV, 61–95.

Farinós, J. 2017a. Gobernanza territorial sin territorio. In Farinós, J.; Serrano, A. (eds.) Ordenación del Territorio, Urbanismo y Medio Ambiente en un mundo en cambio. Valencia: PUV-Cátedra de Cultura Territorial Valenciana, 213–245

Farinós, J. 2017b. Bases conceptuales de la gestión territorial: inteligencia territorial y ética práctica. In Cittadini, E. y Vitale, J. (eds.) Observatorios territoriales para el desarrollo y la sustentabilidad de los territorios. Vol. 1: Marco conceptual y metodológico. Mendoza-San Juan (ARG): Instituto Nacional de Tecnología Agropecuaria, 12–25.

Farinós, J. 2019a. La inteligencia territorial en la toma estratégica de decisiones. In Peñarrubia, M.P. y Simancas, M. (coords.) El valor de los datos turísticos. La transformación de la información en conocimiento para la toma inteligente de decisiones. Valencia/Sta. Cruz de Tenerife: Tirant lo Blanch/ Catedra de Turismo CajaCanarias – ASHOTEL- Universidad de La Laguna, 23–41.

Farinós, J. 2019b: La cooperación horizontal de carácter territorial entre CCAA, un reto para la política de OT y para el modelo de organización del Estado. In Farinós, J.; Ojeda, J. y Trillo, J.M. (eds.) España: Geografías para un Estado posmoderno. Madrid/Barcelona, AGE/Geocrítica, 187–205.

Farinós, J. & González, M. 2019. La gobernanza territorial como concepto, proceso y resultado. In Romero, A. y Alejo, A. (coords.) Gobernanza. Perspectivas y retos para su estudio (en prensa). Valencia: Universidad de Alicante-Tirant Lo Blanch (en prensa).

Farinós, J.; Peiro, E. & Quintanilla, P. 2017. Cultura Territorial: de la información al conocimiento y el compromiso para la acción ciudadana. La iniciativa de la "Cátedra de Cultura Territorial Valenciana. Proyección XI (22), 131–153.

Farinós, J. & VERA, O. 2016. Planificación territorial fronética y ética práctica. Acortando las distancias entre plan y poder (política). Finisterra 101, 51–75.

Flores Ruiz, D.; Perogil Burgos, J. & Miedes Ugarte, B. 2018. ¿Destinos turísticos inteligentes o territorios inteligentes? Estudios de casos en España. Revista de Estudios Regionales 113, 193–219.

Foucault, M. 1982. The subject and power. In Dreyfus, H. & P. Rabinow, P. (eds.) Michel Foucault: Beyond Structuralism and Hermeneutics. Brighton: Harvester Press, 214–232.

Gallego, J.R. & Pitxer, J.V. 2018. Reinterpretando el desarrollo territorial: una nueva visión desde la economía. In Farinós, J. (coord.), Territorio y estados: elementos para la coordinación de las políticas de ordenación del territorio en el siglo XXI. Valencia: Tirant lo Blanch. Tirant Monografías, 189–232.

Gallego, J.R. & Pitxer, J.V. 2019. Propuesta de articulación institucional entre la planificación física y la planificación económica: un enfoque evolucionista y territorial. Actas del IX Congreso Internacional de Ordenación del Territorio, Santander, 12–15 de octubre, 311–327.

Gaucherend, A. 2006. Introduction à la notion d'Intelligence Territoriale, https://docplayer.fr/10478284-Introduction-a-la-notion-d-intelligence-territoriale.html (accessed 03.08.19).

Habermas, J. 1985. Questions and Counterquestions. In Bernstein, R.J. (ed.) Habermas and Modernity. Cambridge, Massachusetts: MIT Press.

Healey, P. 2007. On the Social Nature of Planning. Planning Theory & Practice 8(2), 133–136.

Hooghe, L. & MARKS, G. 2001. Types of multi-level governance. European Integration online Papers (EIoP) Vol. 5 (2001) N° 11. <http://eiop.or.at/eiop/pdf/2001-011.pdf> (accessed 03.08.19).

Inneraty, D. 2010. La gobernanza de los territorios inteligentes. EKONOMIAZ. Revista Vasca de Economía 74 (02), 50–65.

Keating, M. 1997. The invention of regions: political restructuring and territorial government in Western Europe. Environment and planning C 15, 383–398.

Luhmann, N. 1987. SozialeSysteme. GrundrisseinerallgemeinenTheorie. Frankfurt a. M.: Suhrkamp.

Mata, R. 2017. El plan nacional de paisaje cultural: Una iniciativa para el conocimiento, la cooperación y la salvaguarda de paisajes de alto interés cultural. In Serrano, A. (coord.) Farinós, J. y Serrano, A. (eds.) Ordenación del territorio, urbanismo y medio ambiente en un mundo en cambio. Valencia: PUV-Cátedra de Cultura Territorial Valenciana, 287–302.

Marks, G. & Hooghe, L. 2004. Contrasting visions of multilevel governance. In Bache, I. & Flinders, M. (eds.) Multilevel governance. Oxford: Oxford University Press, 15–30.

Montufo, A.M. 2017. La protección del patrimonio territorial, teorías, conceptos, normativas y casos de estudio en Granada. E-rph, 20, 5–56.

Moulaert, F. 2010. Introduction: Challenges for social innovation research. In Moulaert, F.; MacCallum, D.; Mehmood, A. & Hamdouch, A. (eds.) Social Innovation: Collective action, social learning and transdisciplinary research. KATARSIS Contract Nr. 029044 (CIT5), WP5 Methodology Development (D5), and Final Report: Towards a Handbook (D6), 7–10.

Nadin, V. 2006. The Role and Scope of Spatial Planning Literature Review. Spatial Plans in Practice. Supporting the Reform of Spatial Planning. Londres: Department for Communities and Local Government.

Paasi, A. 2003. Region and place: Regional identity in question. Progress in Human Geography 27, 475–85.

Pfeilstetter, R. 2011. El territorio como sistema social autopoiético. Pensando en alternativas teóricas al "espacio administrativo" y a la "comunidad local". perifèria 14, 1–17.

OSE-AAVV. 2009. Patrimonio natural, cultural y paisajístico. Claves para la sostenibilidad territorial. Madrid: Observatorio de la Sostenibilidad en España- Ministerio de Medio Ambiente, Rural y Marino.

Pinson, G. 2011. Urbanismo y gobernanza de las ciudades europeas. Gobernar la ciudad por proyecto. Valencia: IIDL/PUV, Colección 'Desarrollo Territorial' 10.

Pulido-Fernández, MC.; Pulido-Fernández, J. 2019. Is There a Good Model for Implementing Governance in Tourist Destinations? The Opinion of Experts. Sustainability (11)12: 3342.

Raffestin, C. 2011. Por una geografía del poder. El Colegio de Michoacan Traducción y notas Yanga Villagómez Velázquez. (Original in French: Pour une géographie du pouvoir. París: LITEC, 1980).

Romero, J. y Farinós, J. 2011. Redescubriendo la gobernanza más allá del buen gobierno. Democracia como base, desarrollo territorial como resultado. Boletín de la Asociación de Geógrafos Españoles 56, 295–319.

Sami, M. & PAASI, M. 2013. From geopolitical to geoeconomic? The changing political rationalities of state space. Geopolitics 18(2), 267–283.

Sen, A. 1999. Development as Freedom. Oxford: Oxford University Press; New York: Alfred Knopf.

Serrano, A. 2017. Paisaje, patrimonio territorial y la necesidad de nuevas formas y herramientas de planificación territorial y urbana. In Serrano, A. (coord.) Farinós, J. y Serrano, A. (eds.) Ordenación del territorio, urbanismo y medio ambiente en un mundo en cambio. Valencia: PUV-Cátedra de Cultura Territorial Valenciana, 249–286.

Soja, E. W. 1996. Thirdspace. Malden (Massachusetts): Blackwell.

Swyngedouw, E. 2004. Globalisation or 'glocalisation'? Networks, territories and rescaling. Cambridge Review of International Affairs 17(1), 25–48.

Troitiño, M. A. 2008. La extrapolación de los principios del Patrimonio Mundial al Patrimonio Local: los retos de los municipios para garantizar el desarrollo económico sostenible con la gestión del patrimonio cultural. II Taller de Técnicos y Gestores de Patrimonio. Aranjuez.

Troitiño, M. A. 2011. Territorio, patrimonio y paisaje. Ciudad y Territorio. Estudios Territoriales. 169/170, 561–569.

Troitiño, M.A. 2019. El desafío y la necesidad de construir proyectos territoriales en clave patrimonial- Actas del IX Congreso Internacional de Ordenación del Territorio, Santander 12–15 de marzo, 1846–1853.

Troitiño Vinuesa, M.A. & Troitiño Torralba, L. 2018. Visión territorial del patrimonio y sostenibilidad del turismo. Boletín de la Asociación de Geógrafos Españoles 78, 212–244.

UNESCO. 2003. Texto de la Convención para la Salvaguardia del Patrimonio Cultural Inmaterial.

Zapouzaki, K. 2018. Resilience versus public spatial planning in times of crisis: lessons from Greece. In Farinós, J. (coord.) Territorio y estados: elementos para la coordinación de las políticas de ordenación del territorio en el siglo XXI. Valencia: Tirant lo Blanch. Tirant Monografías, 1159–1188.

Perceptions mapping: A participatory tool for urban conservation planning

C. Ost & R. Saleh
ICHEC Brussels Management School, Brussels, Belgium

ABSTRACT: This paper departs from questioning the relationship between the everyday maker and the built environment. It thus, positions the human reflections and daily interactions with the cultural heritage in terms of human sensory experiences at the center of its empirical research. The process of perceptions mapping is a sense-making process during which people map what they feel their cultural, natural and human assets are; express their opinions, ideas, needs and aspirations but also; raise concerns and highlight conflicts related to the management, conservation and preservation of the cultural heritage for future generations. Departing from the perceptions, the collective memory of what a place was to the community arises. Likewise, diverging and/or converging perspectives emerge in reference to what it is today and above all, how the community would like it to be tomorrow.

1 INTRODUCTION

"Moving elements in a city, and in particular the people and their activities, are as important as the stationary physical parts... Most often, our perception of the city is not sustained, but rather partial fragmentary, mixed with other concerns. Nearly every sense is in operation, and the image is the composite of them all" ((Lynch 1960:2).

This article wishes to posit perceptions mapping as a participatory tool for probing the relationship between the everyday maker and the built environment. A tool that positions human preferences, reflections and daily interactions with the cultural heritage in terms of sensorial experiences (hearing, touching, seeing, tasting and smelling), at the center of its empirical research. We would like thus, to postulate perceptions mapping as a sensemaking process (Weick 1995) during which people map their cultural, natural and human assets; express and exchange their opinions, ideas, needs and aspirations but also raise concerns and highlight conflicts related to the management, conservation and preservation of the cultural heritage for future generations.

This article will resume the results of a one-year empirical research during which perceptions mapping was exploited for analyzing and visualizing "attributed values" based on individual/collective memory in relation to the perceived cultural heritage. Likewise, we will discuss the potential of perceptions mapping in capturing the diverging and/or converging perspectives in reference to what the cultural heritage represents today and above all, how some members of the community would like it to be tomorrow. Proceeding from this premise, we would like to put forward perceptions mapping as a two-folded tool; an ex-post

reflection tool and a co-design medium. As an ex-post reflection tool, perceptions mapping demonstrated to be very handy. First, it facilitated the evaluation of previous urban development projects. Secondly, it resuscitated and reinstated on the map some forgotten/erased intangible heritage assets alongside with the cherished tangible assets. Indeed, as per Lynch's (1960:1) statement: *Every citizen has had long associations with some part of his city, and his image is soaked in memories and meanings.* The ex-post evaluation of urban conservation planning is done through a storytelling process during which sensations, feelings, individual and collective memories are materialized on a physical map. While as a co-design tool, perceptions mapping visualized and expressed people's projections and proposals in a participatory, amusing and user-friendly manner thanks to the embracement of the mapping methodology and icons developed by Map-it toolkit (Dreessen et al. 2012)). Although some participants were skeptical about the review and maybe, the potential adoption of the maps by the authorities, the co-created maps are perceived as a tool for exercising agency. Architecture and planning intervene in the definition of daily space. The contribution of both disciplines to the socio-spatial transformation is above all linked to the "*social construction of reality*" (Berger & Luckmann 1966). Very often, in fact, the conceived space is out of the question (Lefebvre 2008), and this is precisely the key to the relationship between the built environment and power. "*The more that the structures and representations of power can be embedded in the framework of everyday life, the less questionable they become and the more effectively they can work. This is what lends built form a prime role as ideology*" ((Dovey 1996:2). At the end of the day "*power comes from maps and it traverses the way maps*

DOI 10.1201/9781003182016-11

are made" (Harley 1989:12). In this regard, this article aims at speculating on how perceptions mapping could become a new tool for assessing the state of the art; measuring people's attributed values; and avoiding "cosmetic" consulting activities. A tool for orienting future urban conservation planning vs co-design.

2 SETTING THE SCENE

2.1 *The empirical research under the framework of H2020 project CLIC*

The perceptions mapping process, was carried out in tandem with a mapping process of the state of the art of the urban development process. The two processes (perceived and *de facto*) are conceived as requirements of the Historic Urban Landscape (HUL) approach which is embedded in this research. Our research was performed under the framework of the H2020 CLIC project[1] and it involved three partner cities/region of the CLIC Consortium, namely: Rijeka in Croatia, Salerno in Italy and Vastra Götaland Region in Sweden.

In order to identify the multilayers and interconnections between the human, natural and cultural (tangible and intangible), international and local values present in our CLIC cities/region, the HUL approach was adopted. The documentation phase of the state of the art was structured in three different moments. Firstly, we asked our CLIC cities/region to provide us with data related to the urban component at the macro level (geological and topographic mapping, environmental mapping, regulation mapping, historical and cartographic mapping, mobility mapping and current land use mapping). Secondly, we asked our CLIC cities/region to provide us with data related to the heritage component at the meso level. In this phase, we agreed together on the boundaries and identification of the cultural heritage. Finally, CLIC cities/region were asked to provide us with data related to the selected sites for adaptive reuse at the micro level (characteristics of the cultural heritage, economic and management aspects, conservation status, potentials and constraints for its reuse, accessibility/ proximity, and existing ideas of adaptive reuse). Thanks to the richness of the collected data by CLIC partners in addition to extra field and desk research conducted by ICHEC's interns, we were able to map not just tangible and intangible heritage but also other cultural and natural assets, and how these are connected and spatially integrated.

In the production of space (1974), Lefebvre underlines the dialectic between the social construction of space and the of the everyday practice. For the French scholar, the concept of space is reflected as a means of production, but also as a product. Lefebvre depicts space as a social product, a means of social reproduction and control. Indeed, he posits that the production of space takes place through three spatial dimensions: "conceived space", "perceived space" and "lived space". The "conceived space" represents the technical language of design and spatial agent. While the "perceived space" provides the materials for the reproduction of a society, based on the daily spatial routines. Finally, the "lived space" is where the imagination tries to change and appropriate itself. According to him, the production of the "lived space" is the result of a struggle between appropriation and dispossession. Our aim here is to negotiate between the conceived (technical blueprint) and the perceived spaces (actual use of space) in order to help people reconnect with their memories and fulfil their desired/imagined space (the lived space).

Since people are an integral part of the city's ecosystem, the mapping state of the art had to be confronted and complemented with people's perceptions. As stated above, the UNESCO Recommendation on Historic Urban Landscape pays meticulous attention to "*perceptions and visual relationships*" and to "*the intangible dimensions of heritage as related to diversity and identity*". Considering that human interactions and sensorial experiences are an integral part of the *genius loci* of the place, perceptions mapping was deemed as a symbiotic part of the research. It is however fascinating to see how Lynch has already anticipated and framed this in the image of the city: "*Moving elements in a city, and in particular the people and their activities, are as important as the stationary physical parts... Most often, our perception of the city is not sustained, but rather partial fragmentary, mixed with other concerns. Nearly every sense is in operation, and the image is the composite of them all*" (Lynch 1960:2).

Our mapping process was focused on people's perceptions, opinions and feelings with regards to their cultural heritage. In designing the participatory methodology, the Faro Convention was also embedded. This convention emphasizes on the value of cultural heritage as assets for sustainable development and a better quality of life. The peculiarity of our choice is because the convention also highlights the importance of the *heritage community* as an empowered community that aspires to conserve and safeguard these common goods for future generations (Council of Europe 2005). Our scope of investigation was to valorize the interactions between the human, and cultural heritage, alias, the values that make our cities unique and characteristic. Whilst the main objective of this mapping process was not only to capture this intimate interrelationship but also to provide a methodology for citizen's participation in evaluating and co-designing urban conservation plans.

During the perceptions mapping workshops, we learned that people's mental maps are composed of

[1] This research has been developed under the framework of Horizon 2020 research project CLIC: Circular models Leveraging Investments in Cultural heritage adaptive reuse. This project has received funding from the European Union's Horizon 2020 research and innovation programme under grant agreement No 776758.

what they see, touch, odor, taste or hear in relation to the cultural heritage. These mental maps represented participants' spatial knowledge and interest in peculiar tangible and intangible assets. Basar describes beautifully what we experienced during the workshops: *"...everyone, whether educated in architecture or not, is affected by and has an effect upon the spaces they occupy: you are born somewhere (a house, a hospital) you live somewhere (a flat, a farm) and you die somewhere (a house, a hospital). Everyone is secretly, profoundly cultured about their built world, they just don't know it, yet"* ((Miessen M. and Basar S., 2006):32). Indeed, people from all walks of life were proactively engaged in describing scrupulously the uniqueness of their lived environment. Participants recalled single and collective memories and evoked foregone and unpredictable elements. At the end of the day, every person was proud and satisfied of his/her contribution.

As this research departs from the theoretical setting of the Historic Urban Landscape approach (HUL), and eventually contributes to the economic politics of the place, it is deemed important to explain such theoretical framework.

2.2 *Historic urban landscape*

In 2011, UNESCO adopted a recommendation on the Historic Urban Landscape (HUL) as a new approach to urban conservation which takes into consideration the interconnections between the multilayered values of the historic city; the human, cultural and natural assets (, Bandarin & van Oers 2012; Bandarin & van Oers 2014; Pereira 2019; UNESCO 2011).

"The historic urban landscape is the urban area understood as the result of a historic layering of cultural and natural values and attributes, extending beyond the notion of 'historic centre' or 'ensemble' to include the broader urban context and its geographical setting. This wider context includes notably the site's topography, geomorphology, hydrology and natural features, its built environment, both historic and contemporary, its infrastructures above and below ground, its open spaces and gardens, its land use patterns and spatial organization, perceptions and visual relationships, as well as all other elements of the urban structure. It also includes social and cultural practices and values, economic processes and the intangible dimensions of heritage as related to diversity and identity" (UNESCO 2011).

This policy and planning tool builds on values related to the human interactions with the built and natural environment as well as communities' perceptions. It thus represents a holistic and integrated vision of the historic city. HUL applies an interdisciplinary investigation of the dynamic and relentlessly changing historic cities. It aims at preserving the integrity of historic, social and artistic values within a sustainable development perspective (Bille 2018l; Throsby 2017; Labadi 2016). According to the HUL approach, the distinctive values, aka the DNA of a place, should be considered as a prelude in the overall management and development of the city. In this sense, the HUL represents a new perspective of understanding the uniqueness of our lived environment. Under this framework, the spatial investigation identifies multidimensional layers through the tailored tools to the local context. This peculiar investigation, constitutes a richness of a breadth and depth that needs to be acknowledged and enhanced in the urban conservation and development plans.

Among such many layers, we could identify individual perspectives that contribute to the final HUL (geomorphology, hydrology, demography, economics, social, environmental, etc). Each of such perspectives suggests a state-of-the-art of the complex and holistic vision of the city, to which the outcomes from the perception mapping are eventually confronted.

As an example, we may assume that the economic perspective of the HUL is not a common linear one (from inputs to outputs), but an implicit circular process within which we reuse past resources, we adapt the reuse of a building into new, sustainable, and inclusive needs and urban uses, and at the end of the day we integrate conservation of cultural heritage within the Sustainable Development Goals.

In an urban context, the cultural heritage is composed of different categories. In order to map it, we first of all defined the boundaries of the spatial analysis. In each of our case-studies we defined three levels of urban analysis: the micro level (the building level), the meso level (the historic center) and the macro level (the entire city/municipality). Practically speaking in our case studies, for meso area, we refer to urban ensembles that include altogether, streets, blocks of buildings and public spaces impregnated with tangible and intangible assets, urban cultural and urban green assets.

The micro level was set by the cities/region when each designated between 1-3 immovable heritage for adaptive reuse. The meso and the macro level were drawn in close cooperation with our partner cities/region.

The mapped perceptions in relation to the cultural heritage took place within the meso level boundaries. Practically speaking, the meso level was represented by the historic centers of Rijeka and Salerno; and of four rural municipalities in the Region of Vastra Götaland in Sweden, namely: Fengersfors, Forsvik, Gustavsfors and Strömsfors.

3 THEORETICAL FRAMEWORK

According to Lynch (1960:8), *"an environmental image maybe analyzed into three components: identity, structure and meaning"*. Our investigation was based mainly on these three criteria with specific connotations related to our purpose of research. More specifically, Lynch linked identity to distinctive objects while we investigated identity in terms of tangible and intangible cultural heritage. We believe that the intrinsic value of these assets is the main attribute that triggers its "imageability". Lynch (1960:9) defines imageability as: *"that quality in a physical object which gives it a high probability of evoking a strong image in any given observe. It is that shape, color, or arrangement which facilitates the making of vividly identified, powerfully structured, highly useful mental images of the environment".*

The second criterion, structure, was defined by Lynch as the *"spatial or pattern relation of the object to the observer and to other objects"* (1960):8). We applied the same concept to the spatial relation between the observer, the identified heritage assets, and other objects. The first relationship was investigated through the five senses. In this regard, participants set forth which sense(s) was related to their daily interaction with the heritage asset(s). While the second correlation was investigated through an evaluation of whether the asset in question represented a weakness, a threat, or an opportunity in relation to the lived environment. Finally, Lynch supposed that the observer attaches a practical or emotional meaning to the object. In this regard, three different assessments took place, firstly; we asked people to identify heritage value. Thus, what the perceived cultural heritage within the meso area meant for them. Secondly, we asked people to assign a color to their city and; lastly, people were asked to identify the most visited and liked routes.

The question related to the color was meant to assess how people were affected by the surrounding environment. By analyzing the interviews and visualizing the data we noticed that people attributed a color to the city in relation to their immediate surrounding environment, more specifically, within the interview area. Many people associated the color of the city to the color of buildings paint and/or stone, or a unique heritage especially in the cities. In the rural areas, the association was made in reference to the natural elements (fauna and flora). In some cases, the color of the city was associated to a feeling, a collective memory or a personal/family history anecdote. Although perceptions varied from one person to another, some dominant colors were recurrent either because of the geographic location of the interviewee during the interview, i.e. overlooking the sea, closeness to the forest, the history of the area, etc…or because of the unique character of the assets in the meso area, i.e. unique color of the stones, richness of industrial heritage, landmarks, distinctive architectural elements, local building materials, etc…

In our three cities/Region, the dominant colors pervaded the space. Blue in the coastal cities of Rijeka and Salerno was all over the place whilst green was prevailing in the four rural municipalities of Vastra Götaland.

Nonetheless, it was interesting to notice that some expressed colors represented the city's changing urban narrative in terms of: new urban developments (positive/negative), demolishing/substituting or a reminder of past collective memories. Color is a distinctive characteristic of a city's identity and it answers Lynch's (1972) question: "what time is this place" by unpacking the historical layers and values.

4 METHODOLOGY

Perceptions mapping was carried out in tandem in four partner building/cities/region of the CLIC project consortium: Rijeka (Croatia), Salerno (Italy), Pakhuis de Zwijger (Amsterdam, the Netherlands), and Vastra Götaland Region (Sweden). In Amsterdam, perceptions were mapped in relation to an industrial heritage building, thus at the micro level. Since Pakhuis de Zwijger is a cultural dialogue platform and not a local authority, as in the case of the other three CLIC partners, the perceptions mapping process revolved around the building and its relationship with the surrounding environment.

As anticipated, before launching the perceptions mapping process in Rijeka, Salerno and Vastra Götaland Region, we defined with our partner CLIC cities/region three levels of urban analysis: micro (building level), meso (historic center level), and macro (city level/region). The following paragraph aims at elucidating the two-phase methodology which was developed and put into practice in order to capture people's perceptions, personal interconnections and sentiments in relation to the cultural heritage.

Four master's students were engaged to undertake an internship in Amsterdam, Rijeka, Salerno and Gothenburg (covering five case-studies in Vastra Götaland Region). Thanks to the support of our partner cities/region and related academic partner, the four collected data related to:

1. people's perceptions about their cultural heritage in the meso area through both random and ad-hoc interviews; and
2. the livability of the four cities/region through personal observations.

In order to articulate the perception mapping process, this study performed a structured review of the literature on Historic Urban Landscape (Bandarin & van Oers 2012; Bandarin & van Oers 2014; Bolici et al. 2017; Pereira 2019; Santander et al. 2018; UNESCO 2011), Cultural capital (Benhamou 2012; Ost 2016; Ost 2019; Throsby 2001; Throsby 2002; Throsby 2017), Cultural mapping (Freitas 2016; Hossain & Barata 2019; Jeannotte 2016; Murray 2017; Savić 2017; Pillai 2014; UNESCO 2009), Sensorial mapping (Dubey et al. 2016; Graezer et al. 2017; Hoekstra 2019), Co-design and participatory mapping (Blake et al. 2017; Dreessen et al. 2012; Gutierrez 2019; Miessen & BasarS 2006; Naik et al. 2013; Reilly et al. 2018; Ringholm & Agger 2019; Salesses et al. 2013; Yoshimurab et al. 2018).

4.1 Perceptions mapping: phase one

In the case of Pakhuis de Zwijger phase one was carried out in June 2018, during the festival WeMakeTheCity[2] at Pakhuis de Zwijger. This mapping focused on people's perception about the building. For this purpose, a citizen dialogue kit was used. This smart toolkit which is developed by Research[x]Design, Department of Architecture of KU Leuven university, was specially tailored by Research[x]Design for our case, and used for polling. This first phase helped understanding people's perceptions about the heritage building and its relationship with the surrounding area. The visitors had to answer a number of questions according to their use and knowledge of the building. Three different categories of visitors were identified and three ad-hoc screens were designed for: frequent visitors, Amsterdam citizens, first-time visitors.

Random and selected interviews were conducted in Rijeka, Salerno and Vastra Götaland Region. The sample aimed at representing people from all walks of life. The choice of random and specialized interviewees was envisioned to capture the diversity of perceptions, feelings and opinions about the quality of the lived environment through the five senses (sight, hearing, taste, touch and smell). Open-ended questions (i.e. When you think of your lived environment, what are the cultural heritage elements (tangible and intangible) that shape the identity of this place and makes it unique?) were embedded in semi-structured questionnaires. The interviews had a duration between 30 minutes to one hour and it was conducted in person. The duration of the interview depended on:

1. the availability of the interviewee;
2. linguistic barriers (interviews were conducted in English. Simultaneous translation was available only in Italy);
3. level of knowledge and willingness to share insights and thoughts.

Moreover, the level of detail depended significantly on the aptitude of the interviewers to trigger interest and enable the communication and dialogue with the interviewee. Although language was a barrier, especially in Salerno, thanks to the meticulous work of our interns a number of interviews were conducted as follows: 15 interviews in Rijeka; 22 interviews in Salerno and 12 interviews in four locations in Vastra Götaland Region. A diversity of insights was captured and the sampling criterion applied to the selection of the interviewees was whether people were residing in and/or working in the meso area (Saleh & Ost 2019).

In their survey about the familiarity of Harvard square (Cambridge, MA), (He et al. 2018), demonstrated that "*the relationship between the spatial structure of the built environment and inhabitant's memory of the city derives from their perceptual knowledge*". Following the analysis of 394 samples (out of 3617) the scholars concluded that "*human activity patterns are the drivers of spatial knowledge, which in turn largely depends on temporal parameters*". According to the authors, people who live and work in a specific area tend to be more familiar with the places. It is the frequency of visits that enriches people's familiarity of places and thus capacity to mentally map them. Indeed, this was the driver behind interviewing people who either lived and/or worked in the meso area. For the sake of this research, we were interested in interacting and establishing an enduring relationship with people knowledgeable about their cultural heritage and capable of describing it. This is also because perceptions mapping was part of a long series of participatory meetings/workshops envisaged by CLIC called Heritage Innovation Partnerships (HIPs) (Garzillo et al. 2018).

4.2 Perceptions mapping: phase two

In Rijeka, Salerno and Vastra Götaland Region, phase two was carried out as a group interaction through a participatory workshop based on active listening, feedback, and reflection. The workshops had the duration of three hours and were conducted in the local language.

Phase one had paved the way towards understanding the urban texture in reference to people's perceptions. More importantly, it facilitated the introducing of contextualized examples during the interactive workshop. While phase two helped identify the cultural assets; threats and weaknesses; and future opportunities in terms of potential adaptive reuse opportunities. Hence, the perceptions mapping process embraced the paradigm shift (demand driven instead of market driven) and it thus, departed from and investigated the urban sustainable needs identified by the everyday makers. As a final result, every group produced a map of how they perceived their cultural heritage.

In the case of Pakhuis de Zwijger, phase two consisted of mapping people's perceptions regarding the impact of Pakhuis de Zwijger as a cultural heritage organization on the community/ies. Especially, the perception of its role as catalyst for participative urban innovation development. For this reason, 25 interviews were carried out with emphasis on the years 1935, 1980 and 2006. Needless to say that people's perceptions changed in each of these turning points of the building's history. A number of interviewees from the local authorities, academics and professional experts, expressed their perceptions in relation to Pakhuis de Zwijger governance module; its impact on the local community; and its future vision. Moreover, the local community was interviewed as well to map perceptions in relation to the role of Pakhuis de Zwijger and its relationship with the surrounding area and the community/ies.

For each of the partner city/region, data were collected thanks to the case-studies provided by the CLIC partner cities/region. Additional data was collected through the qualitative interviews and empirical observations during the one-month internships. Finally, the

---

[2] https://wemakethe.city/en/

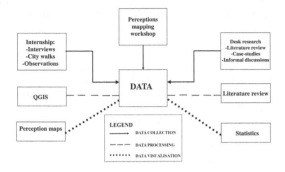

Figure 1. Perceptions mapping methodology. Source: (Saleh & Ost 2019).

interactive workshop provided an unexpected wealth of data and human exchanges. A diverse range of data was collected also through informal conversations with the stakeholders' and civil servants at the workshop, during missions and or CLIC meetings and conferences in the partner cities/region.

The data analysis and elaboration was conducted in six steps and it took into consideration a number of variables:

– The first step was to quantify the perceived elements in every city at the meso level. In tandem, we quantified which sense (sight, hearing, taste, touch and smell) was used more for mapping the elements;
– The second step was to quantify the features that people attached to heritage;
– The third step was to geolocalise the attached color to the place and to draw the routes of the most liked itineraries and places;
– The fourth step was to quantify weaknesses and threats and delimit the bombed areas (conflictual buildings/ areas);
– The fifth step was to quantify the opportunities (how many elements were perceived as not in use or as underused with a potential for future reuse) and;
– The sixth step was to understand how to represent visually the data and how to develop a common but at the same time contextualized legends for Rijeka, Salerno, Fengersfors, Forsvik, Gustavsfors and Strömsfors.

The perceptions mapping workshop was inspired by the five senses methodology which was applied during the HUL workshops in Cuenca, Ecuador. After a thorough analysis of the process and results of this sensory methodology, we decided to adopt it and go beyond. Our added value consists of quantifying data and of visualizing perceptions in a representative and catchy way. We deem this sensory journey relevant because sometimes people feel at ease to express their personal experiences, opinions, thoughts and sentiments through sensations, gestures, observations, personal and collective memories and above all, through face to face interactions and eye contact. Indeed, in Cuenca,

this sensory approach not only captured opinions, feelings and memories related to the lived environment but it has also induced to uttering hidden gems and beauty, potential, discomfort, fears and conflicts (Pérez et al. 2017). This subjective feedback based on personal experiences and knowledge of the territory cannot be apprehended by an external expert or merely through desk research. Indeed, at the end of every perception mapping workshop, every group presented its final presentation. Although this took place at the end of the working day, participants were keen not only at presenting but also to listening to the other presentations and this was very enriching. Despite the fact that people took long time to present, almost every group brought up new issues and thus enhanced the discussion and research. The combination between the feedback from the interviews carried out by our interns and the interactive workshop helped us identify not only the perceived cultural heritage but also weaknesses, threats and opportunities.

In order to facilitate the mapping process. We used Map-it toolkit "*Map-it* is a hands-on tool used to plan, analyze and reconstruct past and future projects and spaces. It is a method to visualize a process in space and time, in a low-tech, open and flexible manner" (Dreessen et al. 2012)). This tool is designed by Luca School of Arts, Belgium, for participatory cartography and conversation. Thanks to this playful tool, sensations were captured and face to face interactions were translated by virtue of the auxiliary icons. More importantly, some people who found it challenging to express articulated dissent in front of the public administration were intrigued by the idea of expressing oneself by colors and icons. Indeed, the bombs stickers were found playful but also powerful as a toll of negotiating power. Indeed, in Rijeka and Salerno some participants opted for the bigger sized bombs. The dissent and conflict utterance was faced differently in Gothenburg. Actually, two participants from Gustavsfors (Vastra Götaland Region) didn't feel at ease to bomb anything. According to them: "the place was too beautiful and no conflicts were to report". Lateral/bilateral discussions proved to be very useful and participants felt safe to express their opinions which resulted in a very collaborative co-creation process even between actors that they don't cooperate in everyday life. For example, in Salerno an impressive number of civil society organizations participated in the perceptions mapping workshop and although some had previous sensitivities they worked harmoniously together. Nonetheless, the unexpected result was that some organizations/associations started working and organizing events and projects together after this experience.

5 RESULTS & DISCUSSION

"In the development of the image, education in seeing will be quite as important as the reshaping of what is

seen. *Indeed, they together form a circular, or hopefully a spiral, process: visual education impelling the citizen to act upon his visual world, and this action causing him to see even more acutely. A highly developed art of urban design is linked to the creation of a critical and attentive audience. If art and audience grow together, then our cities will be a source of daily enjoyment to millions of their inhabitant"* (Lynch 1960:120).

Working in a historical building[3] drenched with intangible memories provided the perfect crib for spurring imagination and stimulus. We also noticed that the use of a physical map contributed to enabling the participants to visualize their ideas, reflect on past and existing practices and design their desired space. Using a playful tool (Map-it) empowered the participants and made them feel at ease. It was interesting to notice how groups and individuals were attributing different values to icons such is like, dislike, danger, bombs, etc…Its worth noticing that after a while participants' felt confident to express oneself with an icon[4].

The heritage elements which were visualized the most were of course the obvious ones and especially those related to the tangible heritage in people's surrounding environment. Nonetheless, Participants were invited to reflect on the intangible assets as well such as annual rituals or cultural events taking place in the streets of the city. For example: carnivals, processions, festivals, artistic manifestations, storytelling and performing arts etc…These hints led people to identify places associated with the collective memory for example: the city beach which is not more accessible in Rijeka[5] and Salerno[6].

A significant added value came from the different age-ranges. As a matter of fact, older participants mapped intangible assets that the younger generation had never heard about. i.e. in Salerno forgotten intangible heritage was listed and verbally described by a couple of elder participants. The same happened in Rijeka for example when we were introduced to the craftsmanship of Kalafati[7]. The Kalafato, or master caulker, was a craftsman employed in shipbuilding and nautical maintenance. Several years of apprenticeship were needed to become a master caulker responsible for caulking ships and wooden boats. An interviewee

spoke about the Rigo janći[8], the cake that represents love and how Morèić, the black figure symbol of the city's carnival represents diversity[9]. Whereas the younger generation mapped current practices that are transforming the perceived space. For example, in Rijeka a participant spoke about an innovative high-tech umbrella designed by a local designer, the Kiša umbrella.

Moreover, in the cases of Rijeka and Salerno, having an external eye, a participant who lives in the city center but not originally from there, contributed to enriching the map with interesting insights and special elements that the locals took for granted. The outsider viewpoint gave a fresh and interesting input and triggered discussions around not previously valued elements.

The mapped intangible assets emphasized the uniqueness of every place. i.e. Rijeka was highlighted as the city of sounds: music, wind (different sounds according to topography), harbour, cranes, traffic, freight train.

The perceptions mapping helped emphasizing on the richness of the cities but also in pinpointing critical issues related to urban conservation and sustainability. In the sense, a lot of emphasis was put on traffic inside the historic centre and the incredible amount of moving/parked cars; architectural barriers; smog; the need of sustainable transport network and more green and open spaces; and the lack of maintenance of historic buildings. Weaknesses and threats were mapped not to denounce mismanagement but instead to use them as leverage to highlight strengths and opportunities. This is exactly where citizens' proposals came into play in order to provide space for pro-activeness and mobilize for collective benefit of the common good.

In Salerno, for a participant of the purple group[10], the map represented a utopian space to be rediscovered. He took a picture of his group work and said: "Salerno che vorrei!" (Salerno which I desire). The reason behind his exclamation is that he listed all the forgotten historical intangible memory and his group bombed the areas that, according to them, impacted negatively the quality of life in the historic centre. Interestingly enough, this sensation was anticipated by Lynch: *A good environmental image gives its possessor an important sense of emotional security. He can establish a harmonious relationship between himself and the outside world* (Lynch 1960:4).

Our main research gap is the lack of control over the representative sample. In our initial plan, a list of requirements related not only to spatial knowledge but also different backgrounds, age-ranges, equal gender representation, number of participants was developed. However, the final sample differed from one place to

---

[3] The three perceptions mapping workshops took place in historical buildings.

[4] It was funny to remark that a participant from Rijeka managed to express her appreciation for receiving chocolate as a nice top up when her energy was starting to fall short with an icon.

[5] Currently it is an off-limit area and property of the Port authority.

[6] Currently there is a commercial harbour there.

[7] Calafati is a very old venetian craftsmanship. The Society for Mutual Assistance of Carpenters and Caulkers – *Società di Mutuo Soccorso fra Carpentieri e Calafati* – was founded in Venice in 1867. Source: https://sanisepo.it/?lang=en visited on 23/08/2019.

[8] https://www.total-croatia-news.com/lifestyle/27596-rigo-jancsi-a-cake-a-legend-a-forbidden-love

[9] http://www.visitrijeka.eu/All_about_Rijeka/Tales_from_Rijeka/Morcic

[10] Participants were divided in groups of 5-7 participants. Each group was assigned a different color.

another. Moreover, due to language barrier, the number of interviewees was restricted. It would have been ideal to reach out to a larger sample in order to represent the pluralistic society in the decision-making process.

6 CONCLUSIONS

Perceptions mapping is a sense-making process during which people map what they feel their cultural, natural and human assets are; express their opinions, ideas, needs and aspirations but also; raise concerns and highlight conflicts related to the management, conservation and preservation of the cultural heritage for future generations.

When we initially asked the partner cities/region whether they had a list of intangible heritage the answer was negative. Nevertheless, despite the small sample size and meager resources, a large inventory of cultural heritage assets, especially intangibles were mapped. Perceptions mapping was applied as an ex-post reflection tool and a co-design medium. At the end of the process, the maps depicted a sensorial journey of people's sense of place. A collective storytelling of the myriad of spatial tangible and intangible identities. Weaknesses and threats were mapped not to denounce mismanagement but instead to use them as leverage to highlight strengths and opportunities. This is exactly where citizens' proposals came into play in order to provide space for pro-activeness and mobilize for collective benefit of the common good. Local citizens need to be engaged and their perceptions about their own heritage count. Indeed, the main idea behind perceptions mapping is to trigger critical reflection about cultural heritage. Different groups in society perceive cultural heritage weaknesses and threats differently and this participatory design process might lead people to start alternative conversations and practices and speculate on the way forward. Perceptions mapping aims at democratising the design process and enabling participation and pitching of creative proposals. It encourages people to partake in the decision making, project planning and evaluation and thus to develop their own sense of community as heritage communities (Council of Europe 2005). The ultimate objective is to design with the people, instead of for the people.

REFERENCES

Bandarin, F. & van Oers, R. 2012. The Historic Urban Landscape: Managing Heritage in an Urban Century, The Historic Urban Landscape: Managing Heritage in an Urban Century. doi: 10.1002/9781119968115.

Bandarin, F. & van Oers, R. 2014. Reconnecting the City: The Historic Urban Landscape Approach and the Future of Urban Heritage, Reconnecting the City: The Historic Urban Landscape Approach and the Future of Urban Heritage. doi: 10.1002/9781118383940.

Benhamou F. 2012. Economie du Patrimoine Culturel. Paris: La Découverte Editions.

Berger P., & Luckmann T. 1966. No TitThe Social Construction of Reality. A Treatise in the Sociology of Knowledgele. USA: Penguin Books.

Bille P. & Larsen, W. L. 2018. World Heritage and Sustainable Development New Directions in World Heritage Management, 1st Edition.

Bolici, R., Gambaro, M. & Giordano, C. 2017. The regaining of public spaces to enhance the historic urban landscape', The Journal of Public Space. doi: 10.5204/jps.v2i1.49.

Council of Europe. 2005. Framework Convention on the Value of Cultural Heritage for Society.

Dovey K. 1999. Framing Places: Mediating Power in Built Form. London & New York: Routledge.

Dreessen, K. Huybrechts, L. Laureyssens, T. Shepers, S. & Baciu, S. 2012. Map-it. A participatory mapping toolkit. Leuven: Acco.

Dubey A., Naik N., Parikh D., & Raskar R., H. C. A. 2016. 'Deep Learning the City: Quantifying Urban Perception At A Global Scale', in Conference paper, 14th European Confer-ence on Computer Vision ECCV 2016. Amsterdam.

Freitas, R. 2016. Cultural mapping as a development tool, City, Culture and Society. doi: 10.1016/j.ccs.2015.10.002.

Garzillo C., Gravagnuolo A., & Ragozino S. 2018. Circular governance models for cultural heritage adaptive reuse: the experimentation of Heritage Innovation Partnerships, Urbanistica informazioni, (278 s.i), pp. 17–23.

Graezer B. F., Pedrazzini Y., Bordone L., & Herrera L., D. P. F. 2017. Mapping controversial memories.

Gutierrez, M. 2019. Maputopias: cartographies of communication, coordination and action—the cases of Ushahidi and InfoAmazonia, GeoJournal. doi: 10.1007/s10708-018-9853-8.

Harley, J. B. 1989. Deconstructing the map, Cartographica. doi: 10.3138/E635-7827-1757-9T53.

He S., Yoshimurab Y., Helferc J., Hackd G., Rattib C., T. N. 2018. Quantifying memories: mapping urban perception.

Hoekstra, M. S. 2019. Creating active citizens? Emotional geographies of citizenship in a diverse and deprived neighbourhood, Environment and Planning C: Politics and Space. doi: 10.1177/2399654418789408.

Hossain, S. & Barata, F. T. 2019. Interpretative mapping in cultural heritage context: Looking at the historic settlement of Khan Jahan in Bangladesh, Journal of Cultural Heritage. doi: 10.1016/j.culher.2018.09.011.

Jeannotte, M. S. 2016. Story-telling about place: Engaging citizens in cultural mapping, City, Culture and Society. doi: 10.1016/j.ccs.2015.07.004.

Lefebvre, H. 2008. The production of space. U.S.A, U.K & Australia: Blackwell Publishing.

Lynch Kevin. 1960. The Image of City, The Image of the City.

Lynch Kevin. 1972. What time is this place. Cambridge, Massa-chusetts and London: The MIT Press.

Miessen M. & Basar S. 2006. The professional amateur, in Did someone say participate. An atlas of spatial practice. Cambridge, Massachusetts: MIT Press, p. 300.

Murray, S. 2017. Creative Cardiff: Utilising cultural mapping for community engagement, City, Culture and Society. doi: 10.1016/j.ccs.2017.08.003.

Naik N., Philipoom J., & Raskar R., H. C. A. 2013. Streetscore - Predicting the Perceived Safety of One Mil-lion Streetscapes. Cambridge, Massachusetts.

Nyseth, T., Ringholm, T. & Agger, A. 2019. Innovative forms of citizen participation at the fringe of the formal planning system, Urban Planning. doi: 10.17645/up.v4i1.1680.

Ost Christian. 2016. Innovative financial approaches for culture in urban development, in UNESCO (ed.) Culture Urban Future, Global Report on Urban Sustainable Development.

Ost Christian. 2019. Urban Economics, in Cody Jeff and Siravo Francesco (ed.) Historic Cities - Issues in Urban Conservation. Los Angeles: Getty Publications.

Pereira Roders, A. & B. F. 2019. Reshaping Urban Conservation. The Historic Urban Landscape Approach in Action. doi: 10.1007/978-981-10-8887-2.

Pérez Rey J., Astudillo S., Siguencia M. & Forero J. 2017. Paisaje Historico Urbano – Historic Urban Landscape. Cuenca. Edited by P. R. Julia. Cuenca, Ecuador: Universidad de Cuenca.

Pillai Janet. 2014. Cultural Mapping, A Guide to Understanding Place, Community and Continuity. Selangor, Malaysia: Strategic Information and Research Development Centre.

Reilly, K., Adamowski, J. & John, K. 2018. Participatory mapping of ecosystem services to understand stakeholders perceptions of the future of the Mactaquac Dam, Canada', Ecosystem Services. doi: 10.1016/j.ecoser.2018.01.002.

Saleh, R. & Ost, C. 2019. Introduction to perceptions mapping: the case of Salerno, Italy, TRIA Territorio della Ricerca su Insediamenti e Ambiente. International journal of urban planning, 12(2), p. 23. doi: 10.6092/2281-4574/6639.

Salesses, P., Schechtner, K. & Hidalgo, C. A. 2013. The Collaborative Image of The City: Mapping the Inequality of Urban Perception, PLoS ONE. doi: 10.1371/journal.pone.0068400.

Santander, A. A., Garai-Olaun, A. A. & Arana, A. de la F. 2018. Historic urban landscapes: A review on trends and methodologies in the urban context of the 21st century, Sustainability (Switzerland). doi: 10.3390/su10082603.

Savić, J. 2017. 'Sense(s) of the city: Cultural mapping in Porto, Portugal', City, Culture and Society. doi: 10.1016/j.ccs.2017.08.001.

Sophia Labadi, W. L. 2016. Urban heritage, development and sustainability.

Throsby, D. 2017. Culturally sustainable development: theoretical concept or practical policy instrument?, International Journal of Cultural Policy. doi: 10.1080/10286632.2017.1280788.

Throsby, D. 2001. Economics and Culture. Cambridge: Cambridge University Press.

Throsby, D. 2002. Cultural Capital and Sustainability Concepts in the Economics of Cultural Heritage, in Della Torre Marta (ed.) Assessing the Values of Cultural Heritage. Los Angeles: The Getty Conservation Institute, pp. 101–117.

UNESCO. 2009. Building Critical Awareness of cultural mapping. A Workshop Facilitation Guide.

UNESCO. 2011. Recommendation on the Historic Urban Landscape, Records of the General Conference - 31st Session.

Weick Karl E. 1995. Sensemaking in organisations. London: Sage. Thousand Oaks, CA: Sage.

The Future of the Past:
Paths towards Participatory Governance for Cultural Heritage – García et al (eds)
© 2021 Taylor & Francis Group, London, ISBN 978-1-032-02129-4

Urban heritage preservation: A new frontier for city governance in Latin America

E. Rojas
Weitzman School of Design, University of Pennsylvania, Philadelphia, USA

ABSTRACT: Concern for the conservation of the urban heritage is no longer the preserve of the cultural elite. The role played by urban heritage in development is increasingly appreciated and local communities, organizations of the civil society, and private citizens are participating in the designation and protection of urban heritage areas. In historic centers private investors are rehabilitating properties to cater for a growing demand from households and business interested in living and trading in urban heritage areas. These trends offer the opportunity to turn the conservation of the urban heritage into a core contributor to attaining more inclusive, safe, resilient and sustainable cities as stipulated in recent international urban development agreements. Implementing this approach poses significant governance challenges. The present work discusses the opportunities emerging from this trend in Latin America and suggests reforms to integrate the traditional urban heritage preservation and urban planning practices.

1 INTRODUCTION

More than fifty year ago the Charter of Venice called the attention of the international community to the need to protect our material heritage. Since then numerous international charters have expanded our understanding of the significance and diversity of the material heritage and contributed to consolidate a canonical approach to its preservation. The focus expanded from monuments of historic, artistic or religious importance to include vernacular architecture, historic urban areas and their intangible heritage (ICOMOS 2020). An integrated understanding of the urban heritage is promoted by UNESCO (2011) in the Recommendation on the Historic Urban Landscape that advocates for the consideration of all its dimensions: intangible, environmental and tangible. The wider role played by the urban heritage in society is also recognized in recent international agreements that highlight its contribution to the social and economic development and environmental sustainability of cities. Today, the urban heritage is increasingly considered a resource for making cities more inclusive, safe, resilient and sustainable (HABITAT 2016). However, putting these concepts into practice poses a significant governance challenge; this document proposes solutions with a focus on cities in Latin America.

2 URBAN HERITAGE'S CONTRIBUTION TO SOCIAL AND ECONOMIC DEVELOPMENT

The heritage provides communities with a variety of socio-cultural and economic services (Throsby 2012)

that are of importance for its development. The immaterial heritage plays a role in the transmission of social conducts—a component of the community's social capital allowing their members to efficiently and securely transact and consume the goods and services that they produce or acquire—and also in expanding the community's human capital with direct impacts on its productivity and innovation capacity (Pérez de Cuéllar 1996). The material heritage makes a multidimensional contribution to the social and cultural development ranging from grounding the community's sense of place to providing physical capital—buildings and public spaces—for conducting social, cultural and productive activities (UNESCO 2016).

In an urban context, the multiple services provided by the heritage are the result of complex interactions among its intangible and material components. This is the case of social practices (like community meetings, festivals or religious celebrations) that take place in public spaces or dedicated buildings and depend on their existence and quality for their continuity and development. Change—a constant feature of cities—have impacts on both forms of urban heritage and can be negative or positive. Negative impacts occur when a public space is taken over by vehicular traffic preventing its use for traditional community gatherings; or when heritage buildings of historic or architectural significance deteriorate due to abandonment or overuse reducing the socio-cultural services they provide to the community. Positive impacts occur when governments conserve monuments or private owners rehabilitate heritage buildings for contemporary uses safeguarding their cultural values.

DOI 10.1201/9781003182016-12

A more general formulation of the above described positive processes would be that cities will sustainably retain their intangible and tangible heritage for the benefit of current and future generations when they use the heritage to serve contemporary needs while retaining its socio-cultural and economic values. Attaining this desirable outcome requires a delicate balance between change and preservation, that is, developing the urban heritage within the limits of its capacity to take on contemporary activities (known as the carrying capacity). The free operation of the real estate markets usually does not ensure this outcome. Private operators in urban markets tend to abandon heritage properties when no short-term profitable uses exist (deterioration) or over-use them if such profitable uses do exist (overutilization). These negative outcomes not only reduce socio-cultural values in urban heritage areas but also represents a lost opportunity for the sustainable development of cities. Unfortunately, governments have limited capacities to correct these market failures. Usually urban heritage preservation regulations focus almost exclusively in the protection of the socio-cultural values and typically by severely restricting change. Additionally, governments do not command sufficient financial and institutional resources to ensure that the usually large and complex urban heritage areas conserve their cultural values while they change and satisfy contemporary needs. What is needed is an approach to urban heritage preservation that actively seeks this equilibrium by mobilizing public and private resources in a context of flexible and consensus-led urban management structures. Adopting this approach—that involves integrating urban heritage conservation and urban development planning (Rodwell 2014)—is a necessary condition to implement the international agreements seeking a wider contribution of the urban heritage to socioeconomic development; it is also a challenging new frontier for the management of cities in Latin America.

Integrating urban heritage preservation and socioeconomic development in cities is more feasible now than a few years ago given the more effective conceptual and implementation tools available to the preservation and urban planning professions. Urban heritage conservators have: a better understanding of the variety of socio-cultural and economic values of the urban heritage (Throsby 2011); more effective methodologies to assess the multidimensional nature of the historic urban landscapes (Bandarin & Van Oers 2012); more clarity about the complementary roles played by the government and the other social actors in the preservation of the heritage (Rojas 2016); and a more generalized acceptance of adaptive rehabilitation as a viable strategy for preserving the cultural significance of the urban heritage while putting it into contemporary use (Bullen & Love 2011). The urban planning profession on its part, have proven strategies and tools to promote and guide the regeneration of urban areas affected by deterioration or overutilization, that is "…to bring about a lasting improvement in

the economic, physical, social and environmental condition of an area that has been, or is, subject to change." (Roberts & Sykes 2008:17). However, further progress is needed in several fronts, notably: methodologies to promote and sustain long-term agreements among social actors concerning the heritage worth preserving and the most effective means to accomplish this objective; methods to design, approve and implement urban development plans and programs that promote the regeneration of urban heritage areas and preserve their cultural values; and put in place effective institutional mechanisms to coordinate the contributions of the growing variety of social actors interested in the preservation of the urban heritage.

Progress in this direction is facilitated by the Historic Urban Landscapes approach (UNESCO 2011) and its derived analytical and planning methods that highlight the interdependences that exist among the cultural, social, environmental and physical dimensions of the urban heritage's contribution to the social and economic development of cities (Bandarin & Van Oers 2012). Studies completed in several cities confirm the analytical advantages of using these methodologies (Rey 2017; Rogers 2017) however, they also highlight the significant institutional developments still required to move the approach into a full operational stage (Ginsarly et al. 2019) the most important being those required to fully integrate heritage preservation into the management of change in cities.

3 THE SUSTAINABLE CONSERVATION OF THE URBAN HERITAGE, A MULTI-DIMENSIONAL GOVERNANCE CHALLENGE

Tacking the issues outlined involves dealing with a complex area of the public realm, namely "…the use of institutions and structures of authority to allocate resources and coordinate and control activities in society", using verbatim Bell's definition of governance (Bell 2002:1). Despite the fact that Latin American countries have institutions and structures of authority for the preservation of the urban heritage and for the management of change in cities, in most cases the resulting governance processes (in Bell's terms) are institutionally fragmentary and use divergent structures of authority ultimately undermining the sustainability of the preservation effort. As described by UNESCO, in Latin America

"Despite the evolution of urban conservation and regeneration practices, they are still perceived as being disconnected from wider urban development issues. Many decision makers still tend to considerer urban conservation as an obstacle to urban development and a deterrent to economic expansion, with the exception of the tourism-based economy. The institutional fragmentation of public and private stakeholders, particularly between urban planning and heritage preservation institutions; the discrepancy between urban regulations and cultural preservation norms; and the

lack of heritage awareness among urban professional tend to emphasize this gap." (UNESCO 2016:122)

Institutional fragmentation in Latin America has its roots in the different origins of the agencies in charge of urban heritage conservation and city management; heritage preservation mostly linked to the administration of cultural affairs and city management to the management of the territory. Each stream of public institution building established its own set of agencies with specific responsibilities, resources and operational procedures.

Institutional fragmentation results from the diversity of cultural institutions responsible for the urban heritage (national ministries of culture, tourism and education; national and local heritage councils and regulatory bodies) and from the multiplicity of institutions involved in city development (national ministries of urban development, infrastructure, economic development and environment; local urban planning offices; public utility companies) leading to ambiguities in the assignment of responsibilities and resources and diverging rules and procedures. The ensuing vertical fragmentation is further complicated by the horizontal institutional fragmentation that commonly occur in the management of metropolitan regions where the urbanized areas extend over the jurisdictional territories of several local and state governments (Rojas 2008).

The fragmentation of the structures of authority on their part have conceptual and operational origins. Regulations by national cultural institutions superimpose (or entirely replace) local urban development regulations and frequently diverge in objectives and intervention mechanisms with the urban development plans adopted by sub-national entities like municipalities and state governments. Most urban heritage preservation regulations seek the protection of the heritage sites by strictly restricting physical and use change. This approach clashes with urban plans and interventions mechanisms oriented to promote the development of urban areas.

There are cases where these problems are addressed and progress towards a more sustainable preservation of the urban heritage is observable. The cases of Quito in Ecuador and Oaxaca in Mexico are exemplary. For extended periods of time they integrated under municipal leadership the regulation, promotion and joint execution of urban heritage preservation and urban development. Also they managed to mobilize the talents and resources of a variety of social actors, public, private and from the civil society (Rojas 2012). These cases provide insights on good governance practices to integrate urban heritage preservation and city development management.

4 URBAN HERITAGE CONSERVATION AS PART OF CITY MANAGEMENT: THE NEW GOVERNANCE FRONTIER

Lessons learned from successful experiences indicate that the effective integration of the urban heritage to the sustainable development of cities requires the *consolidation of institutions* preferably at the local level of government and the use of *flexible structures of authority* framed in a long-term perspective of the contribution of the urban heritage to city development (Rojas 2012). A critical step towards this goal is the re-assignment of responsibilities and resources among different levels of government aiming at ensuring that the institutions in charge of managing urban development also have effective and well-supervised responsibility over the conservation of the urban heritage. Also needed are institutional arrangements capable of attracting and coordinating the contributions of all social actors interested in the conservation efforts. The structures of authority (laws, norms, regulations) on their part must allow change to occur in the urban heritage areas within well-defined limits consistent with the community's conservation objectives. These reforms represent a complex political and technical governance improvement agenda.

4.1 *Devolution of responsibilities and resources: expanding local institutional capacity under national supervision*

In most countries of Latin America, the public responsibilities on matters pertaining the heritage concentrate on cultural agencies of the Central Government that administer laws and resources devoted to the conservation cultural sites and the development of publicly sponsored cultural activities. These arrangements usually work well in matters related to the conservation of the intangible heritage and monuments of national significance, but crumble when applied to urban areas as they move decisions away from the local social actors and prevent change.

The transfer of responsibilities to cities is essential to turn the formulation and implementation of urban heritage preservation regulations into an integral part of the urban development management systems. This will also reduce the conflicts generated by the current system of national regulations overlaying on local regulations that negatively affect local ownership of the conservation process. There is ample room for supervised —that is adhering to national directives—devolution of responsibilities for urban heritage preservation to local governments particularly to larger and institutionally stronger municipalities. In smaller or weaker municipalities, national or regional government can provide technical support in addition to supervision; however, care must be taken to prevent national institutions from capturing local decision-making processes.

Devolution of responsibilities requires resources. Local governments in Latin America—much like in most of the developing world—collect only a small part of the total taxes (just over 10%) and execute only a small proportion of the public expenditures (18%) (Bahl, Linn & Wetzel 2013).These unbalances affect local capacity to discharge new or expanded responsibilities and national governments must act to mitigate them. Local revenue reforms are contentious issues that may take time to resolve but are unavoidable. In the short term, national governments can use direct transfers to enhance local capacity assisting municipalities in fulfilling the national policy objective of preserving the urban heritage. These forms of resource sharing—as long as they are programmed and stable—greatly contribute to expanding the institutional capacity of local governments.

4.2 Long-term planning and flexible preservation regulations: structures of authority to guide change

Managing a city for fast and equitable social and economic development requires sustained proactive interventions guided by a shared vision among social actors. Attaining this goal needs reforming the traditional structures of authority used in most Latin American cities that rely on static and narrowly focused master plans regulating land uses and buildings, and on limited area and sector plans to guide mid-term investment programs. A number of cities have used strategic plans to articulate the community's long-term objectives and guide the formulation and implementation of urban development interventions. However, the implementation of these plans is not always to par with their scope, relevance and legal nature. Changes in the political makeup of local councils and in the agenda of majors often lead to changes in objectives and the abandonment of previous administration's plans and programs. This recurrent shortcoming of local governance requires addressing. Two major reforms would contribute to reduce these problems: greater involvement of all social actors in the preparation and approval of the plans and programs and the establishment of technical institutes to advise local authorities on urban development matters as is done in Curiba, Brazil (Vassoler 2007) or specialized urban development corporations to implement complex long term urban programs as done in Medellin, Colombia (Echeverri & Orsini 2011). Stable support and policy guidance from higher tiers of government can also help the continuity of local policies and plans. In Latin America this has been the case of sanitation and affordable housing where resources and technical advice from the central government agencies allowed local governments to achieve good results (UCLG 2019).

To fully integrate the conservation of the urban heritage into a development-oriented approach to city management the structures of authority used in heritage conservation also need reform. Traditional conservation regulations seek to preserve all the characteristics that give the heritage buildings, public spaces or urban structures their cultural significance; usually aiming at preserving the urban heritage as it was at a given point in history by severely restricting change. This impose significant financial burdens on public sector institutions and economic loses on property owners that cannot easily adapt their heritage properties to contemporary uses. The drawbacks of this approach generate the already mentioned resistance to conservation found in many cities. An option that deserves wider consideration is the gradation of the preservation restrictions based on the cultural significance of the urban heritage. This approach divides the urban heritage according to preferred levels of protection (ranging from the full preservation of monuments of utmost cultural importance to allowing the adaptive rehabilitation of buildings of typological or contextual heritage value)—allowing flexibility to adapt the urban heritage to contemporary uses.

5 REACHING AGREEMENTS: CONFLICT AND COOPERATION IN URBAN HERITAGE AREAS

Integrating urban heritage preservation in the management of urban development is a complex political and technical operation. To succeed it requires the firm support and cooperation of a wide variety of social actors. Strong agreements on the role played by urban heritage in the development of cities are needed to overcome the persistent opposition to conservation by 'development-first' advocates and the 'conservation au trance' preference of some promoters of heritage preservation. Furthermore, the adoption of complex and costly conservation tools—that may range from government-financed full protection of monuments to government-supervised adaptive rehabilitation by private actors—also requires high levels of agreement among the social actors.

5.1 The diversity of social actors

A first step in the path to building the common ground needed for the integration process discussed in this work is to better understand the motivations, expectations and willingness to engage in the process of the social actors involved. Positive attitudes are more likely to emerge if the social actors are included in the decisions and they see mutual benefits in implementing them; however, they are a varied group and not equally motivated and willing to participate in the preservation of the urban heritage. Elsewhere (Rojas 2019) I argue that their generic treatment as "stakeholders" is insufficient to understand their motivations and suggest differentiating them according to the 'significance' they assign to the issue, a point made by Schmitter (2000) when discussing participation processes in governance. Further development of this suggestions indicates that the study of the actors should differentiate them according to how they understand that the urban heritage affects their daily life and how preservation is connected to issues that are of importance to them. This knowledge will clarify the social actors standing concerning key topic for the sustainable conservation of the urban heritage, including: the mix of socio-cultural values they consider worth preserving; the level of preservation to attain; the support (or opposition) they will offer to the implementation of public interventions; or their willingness to cooperate with other social actors to implement preservation actions.

One group would be social actors willing to support the preservation of the urban heritage and able to influence the process in different ways. This group includes those that have the capacity to elect the officials that manage public affairs in the urban heritage area; they can be called 'right holders' in consideration to the fact that through their voting powers they exercise their capacity to influence the allocation of public resources and the use of preservation regulations affecting property rights. The group directly responding to the desires of voters are the elected officials of the national, state, and local governments, a group that Schmitter (2000) calls 'status holders' in consideration to the fact that they act in representation of others. There are social actors like philanthropists that would be interested in preserving the urban heritage to fulfil their social responsibility commitments; they are a form of 'special interest holders'. Other types of special interest holders are members of the urban community residing in the heritage area and driven to support its preservation acting in defense of their ways of living, a stance documented by Uribe (2014) in the cases of the Yungay and Viel neighborhoods in Santiago, Chile. There are individuals or institutions—ranging from academic entities studying the aesthetic value of buildings to individuals holding the collective memory of a community—that are the 'knowledge holders.'

A second group would be made by social actors benefiting or loosing from the preservation of the urban heritage. This is the case of those affected in their daily life by the conservation or loss of urban heritage, a group properly considered 'stake holders'. Among them there are residents, merchants and craft persons that may gain from preservation interventions that bring better quality of life, more customers or higher property values. Others—particularly renters—may lose when heritage preservation leads to gentrification in the urban area where they live and trade, and they are eventually displaced by the disappearance of affordable rentals. Another set of interest holder is made by the property owners that may experience economic loses as a result of the strict restrictions imposed by conservation regulations or stand to gain in cases where the government conservation programs increase the attractiveness of an urban heritage area. They can be considered 'share holders' as their wealth is affected by the conservation preferences of the rest of society. A significant group of interest holders is made by those that receive temporal benefits from the preserved heritage, a class called by Schmitter (2000) 'space holders'. They include persons that go to the area to access government or private services, cultural activities, shop or trade. Some have a relatively permanent interest, for instance, street vendors and artists that make a living there while others only a passing interest, like the tourists that visit the heritage sites only for short periods of time.

5.2 Reaching agreement: community involvement in conserving the urban heritage

Cooperation among social actors does not come easy; their multiplicity makes it difficult to reach agreements on long term issues including a shared vision for the long-term development of the city and the role of urban heritage in achieving the vision. These difficulties also affect the implementation of agreed policies and programs due to the divisions and dissent that normally arise when allocating scarce resources to competing demands.

In a given urban heritage area there would be social actors adopting a very pro-active attitude willing to commit resources to conservation interventions and others with no interest at all. Social actors in the first group (although with wide variations among countries and cities) usually include: the government (frequently the central government, more recently city governments) that cares for the value of the heritage as a bequest to future generations; committed taxpayers and voters that support the allocation of public resources to the task (commonly scholars, members of the cultural elite, community associations); and individual philanthropists or philanthropic institutions interested in preservation either for its cultural and social significance or for the public relations benefits that can be derived from contributions to community-relevant goals. Several social actors—individuals or institutions, private or semi-private—would be willing to cooperate with others contributing knowledge, financing, and political support and would be prepared to harmonize their urban heritage-related activities with other actors, either formally (through institutions) or informally (through voluntary agreements). There will always be some social actors that will be indifferent to urban heritage conservation. One type commonly found in Latin American cities are non-resident 'right holders' (voters that elect local, regional or national officials) that are unaware or do not value the socio-cultural benefits provided by the urban heritage area and are unwilling to support the allocation of national or local funds to conservation. The growing international and local appreciation of the multiple contributions of urban heritage to development suggests that this latter group would soon be a minority.

6 THE CONVERGENCE OF URBAN PLANNING AND HISTORIC PRESERVATION METHODOLOGIES AND TOOLS

Managing change in urban heritage areas confront decision makers with the need to attain a delicate balance in allowing the urban development to take place while preserving the cultural values of the heritage. Planning methodologies and legally binding regulatory instruments promoting this type of balance do exist but are not widely used nor integrated into urban governance in Latin America. Furthermore, institutional arrangements to foster cooperation among public, private and community-based social actors are also lacking in most cities even though the successful cases already mentioned suggest that they work and that can be implemented in Latin America.

6.1 Planning and fulfilling a shared future

Many cities in Latin America used the Strategic Planning methodology to define long-term goals, identify urban development programs to bring them to fruition and regularly update them to adjust to changing circumstances (CIDEU 2020). The process leading to agreements among the wide diversity of urban social actors called to participate is driven by the use of sound scientifically based knowledge and participatory envisioning methodologies supported by a permanent technical advisory team usually housed on a local academic institution. Translation of these agreements into action is the responsibility of the local governments thus the methodology tends to achieve better results in decentralized urban governance contexts. Ensuring that the envisioned future materialize often requires strong local interventions with solid community support. This is the case of cities fighting the ill-effects of excess of tourism like Barcelona with the introduction of tight regulations to short-term housing rentals (Ayuntamiento de Barcelona 2017) and Amsterdam controlling the use of the city center by visitors (Municipality of Amsterdam 2018).

The capacity to promote agreements on a vision for the future of a city and sustaining it during the implementation of complex urban interventions is one of the strengths of the Negotiated Urbanism, an approach used in France (Christy 2016). The key feature of this approach is the continuous involvement of the community in the implementation of the consensus agreed vision for the future for an urban area. Two conditions are needed for the successful implementation of this approach. The first involves structured consultations with interest holders from different areas of the city (neighborhood, district, city and metro region) and different sectors (commerce, industry, crafts, culture) to reach agreements on what they want from the urban area under consideration. The objectives agreed among the interest holders become the planning brief for the urban specialists preparing the plans. The second condition is the regular reporting to the community by elected officials and the urban specialists on the implementation of the plans. The dual consultation process promotes the sustained community support for the plan and transparency and accountability in the implementation process. The approach—successfully used by the metropolitan region of Bordeaux in the rehabilitation and redevelopment of the old port area—requires strong leadership from the local government and extensive consultation with interest holders, an institutionally demanding and time-consuming operation. Certainly, the cultural significance of many urban heritage areas deserves such concern and resources to ensure long-term success and support.

6.2 Flexible conservation regulations for rapidly changing cities

The implementation of a graded and more flexible approach to conservation also poses significant governance challenges in particular the use of structures of authority allowing flexible conservation regulations. Using the conceptual framework of the building typologies analysis (Caniggia y Maffei 2001) the Municipality of Cartagena in Colombia developed regulations to manage change in the adaptive rehabilitation of private heritage buildings in the Historic Center, a UNESCO World Heritage (Alcaldía Mayor de Cartagena de Indias, 2001). The regulations specify the allowed transformations and uses for each type of residential buildings and are administered by urban curators—private conservators licensed by the Municipality—although the final approval of the projects rest in the Planning Department of the Municipality. The regulations have worked well since their adoption in the late 1990s allowing the adaptive rehabilitation of many buildings for residential, commercial, tourism and recreational uses. However, in the last 10 years high development pressures originating in the rapid growth of mass tourism brought about interventions to residential buildings that do not fully comply with the spirit of the Regulations and banalizing the rich Colonial and Afro-Colombian heritage of the historic center creating conflicts within the city, between the city and the national government, and with UNESCO's World Heritage Center. The difficulties encountered by Cartagena do not invalidate the flexible approach to regulating change in urban heritage areas but stresses the need for high levels of involvement of the community in the implementation of flexible structures of authority for heritage preservation and the continuous supervision of the national institutions, particularly in World Heritage Sites.

6.3 Sharing the burden: public, private and people partnerships

Complex urban operations require complex institutional implementation structures; the conservation of the urban heritage areas—among the most complex urban operations confronted by a city—do need them. The successful implementation of conservation interventions requires significant levels of coordination and resource sharing among public actors, between private and public actors, and the full involvement of the interested communities.

Public-public cooperation is the first necessary condition. Getting the multitude of public agencies involved in urban development and in the preservation of the urban heritage to operate together is a significant challenge. Cities with strong municipalities often opt for establishing public corporations to implement urban operations or manage public services.

In Latin America there are successful examples of these institutional arrangements. A well-known example centered in the development of underutilized public lands containing heritage buildings in the center of Buenos Aires, Argentina, is the Puerto Madero Corporation (Garay et al. 2013). The Public Enterprises of Medellin in Colombia is an outstanding Latin American example of a multi-purpose municipal public utility company (Varela 2009). Municipal Development Corporations are common in institutionally strong municipalities in Brazil, Colombia and Ecuador; they are open ended and multipurpose, and—with adequate supervision and accountability—are effective institutions to implement complex urban operations. Some Latin America countries have legislation enabling public agencies to enter into temporal multi entity contracts (Contract-Plans) to coordinate their contributions to the implementation of a multi sector project. This public-public coordination mechanism—a flexible implementation tool commonly used in France (Liu 2018)—is less commonly used in Latin American where decision makers are concerned of not being able to meet the mid-term budgetary and operational commitments that they entail.

There are many examples of successful public-private coordination arrangements in implementing complex urban operations, some in the conservation of the urban heritage. In the 1990s Quito in Ecuador used a mixed-capital (public and private) corporation to rehabilitate the historic center (Fox et al. 2005). Other forms of cooperation are also common although restricted to monuments. They include private sponsorships for the conservation of monuments and the use of concessions agreements to allow private entities the use of public monuments in exchange for their rehabilitation, operation and maintenance and the payment of leasing fees to the public institutions in charge of the property.

Less common in Latin America are public-private-community (people) partnerships although these type of coordination mechanisms can play a significant role in involving the local interest holders in the decisions and implementation of heritage neighborhood conservation projects. Private-private coordination mechanisms in Latin America are limited to a few private foundations specifically supporting urban heritage preservation and the land use and building regulations contained in the urban master plans and special preservation plans.

7 LOCAL COOPERATION AND SHARED RESPONSIBILITY: THE PATH FORWARD

There is a growing agreement about the significant contribution that urban heritage can make to the social and economic development of cities and recent international agreements crystalize the widespread support for these ideas. There are growing calls for taking advantage of these opportunities without delays. There is sufficient knowledge and experience to move in the desired direction. However, the existing urban governance mechanisms (institutional arrangements and structures of authority) in Latin America prevent progress towards this goal. Fractured institutional arrangements at different levels of government based on agencies with silo-style cultures fail to cooperate and engage with the communities. Dated and well-entrenched top-down and regulation-based city planning and heritage preservation management traditions do not promote pro-active attitudes in the pursuit long term objectives nor flexibility in decision-making to accommodate and promote change.

Making progress in the implementation of the international agreements calling to size the development opportunities offered by well-preserved urban heritage areas requires significant reforms to the prevalent city management and urban heritage preservation approaches used in Latin America. The objective shall be attaining a balance between promoting change in cities and preserving the cultural values of the urban heritage and making this equilibrium sustainable in the long term. This is possible with the knowledge and the public and private tools available. However, to make full use of the resources at hand three major reforms are needed. First, transfer the responsibility for the urban heritage from national culture-oriented institutions to local urban development agencies so the decisions about the balance between development and preservation are fundamentally made at the local level, albeit in a transparent way and fully coordinated with, and supervised by, competent higher tier government agencies. Second, make urban heritage preservation regulations and interventions an integral part of urban development plans and city management procedures. Local control of the processes should reduce conflicts and ensure more buy in on the part of the social actors that are more likely to benefit and also pay the costs of urban heritage preservation. Third, incorporate all interested social actors into the urban heritage preservation governance mechanisms by establishing institutional arrangements that promote cooperation and shared responsibility among this diverse set. Cooperation will help finding a common ground on the socially acceptable balance between development and preservation; shared responsibility will also bring more resources to the enterprise better ensuring its long-term sustainability.

These reforms are not easy to implement. In most countries in Latin America they will cut across long established bureaucratic traditions and deal with long-established mistrusts and conflicts among social actors. However, they are urgently needed to implement what is becoming a widely held consensus about how to put the rich urban heritage of Latin America to use for the well-being of the population thus ensuring a future for this inherence from the past.

REFERENCES

Alcaldía Mayor de Cartagena de Indias. 2001. "Decreto No. 0977 de 2001, Plan de Ordenamiento Territorial del

Distrito Turístico y Cultural de Cartagena de Indias (POT)" Octava Parte. Pages 135–220 http://curaduria2cartagena. com/pdf/POT.pdf

Ayuntamiento de Barcelona. 2017. *Plan Especial Urbanístico de Alojamiento Turístico* Barcelona. https://ajuntament.barcelona.cat/pla-allotjaments-turistics/es/

Bahl, R., J. Linn & D. Wetzel. 2013. *Financing Metropolitan Governments in Developing Countries.* Cambridge, MA, Lincoln Institute of Land Policy

Bandarin, F. & R. van Oers. 2012. *The Historic Landscape. Managing Heritage in an Urban Century.* Chichester: Wiley-Blackwell.

Bell, S. (editor). 2002. *Economic Governance and Institutional Dynamics.* Melbourne: Oxford University Press.

Bullen, P.A. & E.D. Love. 2011. "Adaptive Reuse of Heritage Buildings" in *Structural Survey* **29:11**

Caniggia, J.F. & G.L. Maffei. 2001. *Architectural Composition and Building Typology. Interpreting Basic Building.* Florence: Alinea.

Centro Iberoamericano de Desarrollo Estratégico Urbano CIDEU. 2020. Barcelona, Spain https://www.cideu.org Last visited April 11, 2020

Christy, H. 2016. *L'urbanisme Négocié* Paris, Éditions La Découverte

Echeverri, A. & F. Orsini. 2011. "Informalidad y Urbanismo Social en Medellín" *Sostenible* **12** 11–24

Fox, C. With J. Brakarz, & A. Cruz. 2005. *Tripartite Partnerships: Recognizing the Third Sector, Case Study of Urban Rehabilitation in Latin America* Washington, D.C. Inter-American Development Bank https://publications.iadb.org/publications/spanish/document/Alianzas-tripartitas-Reconocimiento-del-tercer-sector-Cinco-estudios-de-casos-en-la-revitalización-urbana-de-América-Latina.pdf

Garay, A., L. Wainer, H. Henderson, & D. Rotbart. 2013. "Puerto Madero: A Critique." *Land Lines* July 2013

Ginzarly, M., C. Houbart, & J. Teller. 2019. "The Historic Urban Landscape approach to urban management: a systematic review," *International Journal of Heritage Studies* **25:10**

International Council on Monuments and Sites ICOMOS. 2020. *Charters Adopted by the General Assembly of ICOMOS* Charenton-le-Pont, France https://www.icomos.org/en/resources/charters-and-texts Last visited April 1, 2020

Liu, J. 2018. Coordination through Integration: A Critical Review on the Spatial Policy and Spatial Planning System of France *International Review for Spatial Planning and Sustainable Development A: Planning Strategies and Design Concepts* **6:3** 125–140

Municipality of Amsterdam. 2018. *City in Balance 2018–2022: Towards a New Equilibrium Between Quality of Life and Hospitality* Amsterdam, Municipality https://www.amsterdam.nl/en/policy/policy-city-balance/

Pérez de Cuéllar, J. 1996. *Our Creative Diversity. Report of the World Commission on Culture and Development.* Paris and Oxford: UNESCO/Oxford & IBH Publishing,

Rey, J. 2017. *Historic Urban Landscape. The Application of the Recommendation on the Historic Urban Landscape in Cuenca, Ecuador.* Cuenca, Universidad de Cuenca-WHTRAP-CPM-MECS The Netherlands.

Roberts, P. & H. Sykes. 2008. *Urban Regeneration: A Handbook* London, Sage

Rodwell, D. 2014. "Celebrating Continuity" *Context* **136:54**

Rogers, A.P. 2017. "Historic Urban Landscape Approach and Living Heritage" in *Conserving Living Urban Heritage: Theoretical Considerations of Continuity and Change.* Cambridge, Cambridge Scholars Publishing

Rojas, E. 2008. "The Metropolitan Regions of Latin America: Problems of Governance and Development" in Rojas, E., J.R. Cuadrado-Roura and J.M. Fernandez (editors) *Governing the Metropolis.* Washington, D.C. Inter-American Development Bank, Harvard University, Rockefeller Center for Latin American Studies. https://publications.iadb.org/en/governing-metropolis-principles-and-cases

Rojas, E. 2012. "Governance in Historic City Core Regeneration Projects" in Licciardi, G. and R. Amirtahmasebi (editors) *The Economics of Uniqueness: Investing in Historic Cores and Cultural Heritage Assets for Sustainable Development.* Washington, DC: The World Bank.

Rojas, E. 2016. "The Sustainable Conservation of the Urban Heritage: A Concern of All Social Actors" in Labadi, S. and W. Logan (editors) *Urban Heritage, Development and Sustainability: International Frameworks, National and Local Governance.* Abingdon Oxon, Routledge

Rojas, E. 2019. "Social actors in urban heritage conservation: do we know enough? In Abrami, E. (editor) *Urban Heritage, Sustainability and Social Inclusion.* New York: Columbia University

Schmitter, P. 2000. Paper presented at the Conference 'Democratic and Participatory Governance: From Citizens to "Holders"', European University Institute, Florence, September.

Throsby, D. 2012. "Heritage Economics: A Conceptual Approach" in Licciardi, G. and R. Amirtahmasebi (editors) *The Economics of Uniqueness: Investing in Historic Cores and Cultural Heritage Assets for Sustainable Development.* Washington, DC: The World Bank.

United Cities and Local Governments UCLG. 2019. *Rethinking Housing Policies: Harnessing Local. Innovations to Address the Global Housing Crisis.* Barcelona, United Cities and Local Governments.

UN Habitat. 2016. *New Urban Agenda,* Nairobi, United Nations Human Settlements Program http://habitat3.org/wp-content/uploads/NUA-English.pdf

United Nations Educational, Scientific and Cultural Organization UNESCO. 2011. "UNESCO Recommendation on the Historic Urban Landscape". Paris, UNESCO

United Nations Educational, Scientific and Cultural Organization UNESCO. 2016. *Culture: Urban Future. Global Report on Culture for Sustainable Urban Development.* Paris, United Nations Educational, Scientific and Cultural Organization. https://unesdoc.unesco.org/ark:/48223/pf0000245999

Uribe, N.2014. Patrimonialización comunitaria en barrios de Santiago: Los casos de las zonas típicas de Viel y Yungay *Apuntes* **17:1**

Varela, E. 2009. Estrategias de expansión y modos de gestión en EPM Medellín. *Estudios Políticos,* Universidad de Antioquia **35** 141–165

Vassoler, I. 2007. *Urban Brazil: Visions, Afflictions, and Governance Lessons.* New York, Cambria Press

The Future of the Past:
Paths towards Participatory Governance for Cultural Heritage – García et al (eds)
© 2021 Taylor & Francis Group, London, ISBN 978-1-032-02129-4

Planning and heritage integration in multilevel governance: Cuenca, Ecuador

M. Siguencia

City Preservation Management Project, Faculty of Architecture, University of Cuenca, Cuenca, Ecuador

ABSTRACT: While in the 90's strategic became increasingly crucial for Spatial Planning, at the same time in the heritage agenda, the concept of Cultural Landscape was established to broaden the span of values and assets. Since then, during the last decades, planning has adopted methods depending on frameworks, demands and contexts. Heritage, meanwhile, has increasingly paid attention to heritage values in urban areas through the development of theories and practices. In 2011, UNESCO adopted the Recommendation on Historic Urban Landscapes, where heritage is a crucial element for sustainable cities' development. In this sense, both Strategic Planning and Heritage Management have shifted, not necessarily 'evolved' towards methods where a common aspect is the crucial role of actors involved. Through the example of the city of Cuenca in Ecuador, this paper discusses the integration of heritage in the city's planning and the involvement of different actors within the governance system.

1 INTRODUCTION

A critical element of the New Urban Agenda 2030 adopted in October 2016 during the United Nations Conference on Housing and Sustainable Urban Development (Habitat III) was the fifteen-year Sustainable Development Goals "SDGs." The essential reflection consisted of the ways of building, managing and living in cities with a significant span of stakeholders or actors that interact at different levels of governance. The SDGs include a specific heritage-driven target in goal 11 numeral 4, "making cities and human settlements inclusive, safe, resilient and sustainable by strengthening efforts to protect and safeguard the world's cultural and natural heritage" (UN Habitat 2016). The inferred close relationship between urban development and heritage is not only a contemporary discourse but traces back to the conception of cities and the recognition of their embedded heritage values.

During the '90s, both in the fields of planning and heritage, concepts advanced. In the field of planning, strategic complements previous rational planning practices; meanwhile, the recognition of a broader set of heritage values under the category of Cultural Landscapes emerges on the heritage agenda. The definition of heritage evolves from object-based to value-based conservation, and more recently, it is moving towards a person-oriented approach (Vandesande 2017). Currently, and as stated in the 2030 Agenda, both fields have sensibly recognised as crucial a focus on people's involvement in planning and heritage management processes. Therefore, decision-making can result in a difficult task when actors have possible conflicts, interests and when the resources oriented to planning and heritage are disjointed. The Recommendation on Historic Urban Landscapes adopted in 2011 by UNESCO incorporates tools (civic engagement, planning, regulatory and financial) to tackle an increasing deterioration or urban heritage areas.

In this context, the present work intends to present the set of stakeholders or actors involved in the current planning and heritage management systems in the city of Cuenca in Ecuador and how they are involved in decision-making processes. In order to do so, this article explores the empirical multi-level governance model based on the work of Pradel Miquel et al. (2013). This model allows a better comprehension of the level in which different groups of actors can influence or transform the institutional framework of policymaking at the city level (Pradel Miquel et al. 2013). Then, the actors in the different levels of governance are distinguished in an institutional and non-institutional framework identifying those related to planning and heritage protection in Cuenca. Finally, an analysis of the integration of both fields is discussed. This paper aims to explore how the integration of heritage within planning systems can be achieved through proper participation and actors' involvement schemes. This study also distinguishes the difficulty of moving from global or national governance to the local, addressing the gap between hegemonic discourse and practice.

DOI 10.1201/9781003182016-13

2.1 *Multi-level governance for planning and heritage*

Strategic Planning provides spaces for actors not only in a consultative role but giving them the voice of what they can do into a development planning program. This singularity recalls democracy and joint agreements linked to social participation. According to Steinberg (2005), the adoption of Strategic Planning tools answers precisely to the search for democracy in planning processes after the longstanding tradition of planning made by urban planners and experts.

In Europe, cities as Hannover in Germany or the Flanders Region in Belgium are examples showing how Strategic Planning addresses diversity with flexible methods accepting the legitimacy of others in a spatial and inclusive environment (Albrechts et al. 2003; Albrechts 2004). Strategic Planning provides tools to produce strategies for cities, regions and sub-regions in Europe involving the construction of new institutional arenas relying on structures of government that are themselves changing (Albrechts et al. 2003). These governance changes aim to promote internal institutional transformations to improve the relation between actors involved, and one of these approaches refers to multi-level governance.

Multi-level governance is an empirical concept based on the distribution of power between the different levels of government and the conception of policymaking partnerships that, besides the state, integrate a multiple set of actors (Pradel Miquel et al. 2013). While this multi-layered system requires transformation on actual governance, the organisation of different actors opens the floor to non-state market and civil society actors. These new forms of state organisation imply an antagonism to the traditional top-down approaches and encourage democratic strategies from civil societies and institutional changes towards social justice (Pradel Miquel et al. 2013).

In Latin America, the concepts of strategic, multi-sectoral and multi-annual planning attract not only planners but also many urban actors considering this as a democratic process in which key actors can reach a standard agreement (Steinberg 2005). This democratic phenomenon has also enticed Latin American countries, which have started to adapt to Strategic Planning. This fact is evident in public-based experiences as the cases of Mexico City, Sao Paulo, and Buenos Aires; partially public as in Lima and Bogotá; and public-private cooperations for strategic planning such as the cases of Guayaquil, San Salvador, and Santiago (Carrión 2014). Although these experiences show an attempt to apply Strategic Planning instruments, especially promoting institutionalisation, there is a lack of legal instruments and management tools.

The differences shown between experiences in Europe and Latin America fall into the sensible context-wise references. For Jajamovich (2013), in Latin America, urban theory keeps following and adapting theories developed in economic and political contexts as different as Europe and the US (Jajamovich 2013), where differentiation between mechanisms, changes, and causes should be first established. For this reason, the context holds historic place-specific characteristics, and therefore strategies cannot be replicated elsewhere (Degen & García 2012; Metzger & Schmitt 2012).

Not only planning, but heritage is a conception strongly linked to its place. The site, at the level of its country, city, or the point on earth's surface, is "sacralised" by its ascribed heritage associations (Ashworth 1994). Even when heritage values can be found at different levels (objects, monuments, cities, landscapes, regions, amongst others), and is recognised in a different manner depending on a particular group of stakeholders, the international regulatory framework delegates the competences for heritage management to State Parties attached to the 1972 Convention concerning the Protection of the World Cultural and Natural Heritage.

State Parties attached to this convention are called to nominate properties on the World Heritage List (WHL). This list includes assets bearing an Outstanding Universal Value (OUV) of global concern and must be managed to maintain this recognition and safeguard properties for future generations. The *Operational Guidelines* are the instrument aiming to facilitate the implementation of the 1972 UNESCO Convention and rules the necessary procedures for the protection and conservation of world heritage properties (UNESCO 2015). Regarding heritage management, the guidelines encourage the establishment of national or regional centres as well as integrate heritage protection into comprehensive planning programs respecting the sovereignty of the State Parties.

Hence, in the heritage filed, levels of governance can also be identified. Culture and arts have increasingly played an essential role in local development strategies in the last decades (Andre et al. 2013). Since cultural assets are not only material but immaterial expressions, they are sources of social identity and interaction between members of a group (Zukin 2012). Different groups aware of heritage have historically organised themselves as governmental and non-governmental entities. As noted by Murzyn-Kupisz & Działek (2013), "non-governmental organisations can constitute a tool of control of authorities and stimulate political participation of citizens; however, they would be less valuable without family or neighbour ties."

Limited experiences have jointly displayed heritage integrated into planning through reforms in governance with diverse actors' interaction. Moulaert et al. (2007) has contributed to the concept of social innovation recognising the role of arts and culture to improve communication within communities. There is, therefore, a recalled need for social involvement since different social organisations intervene and propose new variations upon the outset of certain regions, not necessarily within the governance system (Metzger & Schmitt 2012).

In the concept of social innovation, the re-appropriation of social spaces and the exciting process of flexible institutionalisation give an important political role to associations as public actors on the national and international scene based on the self-management of collective services (Moulaert et al. 2007). Cultural heritage can be considered as a collective construction that may be managed through society's appropriation.

The abovementioned *Operational Guidelines* encourage to ensure the participation of a wide variety of stakeholders, including site managers, local and regional governments, local communities, Non-Governmental Organisations (NGOs), and other interested partners in the identification, nomination, and protection of heritage (UNESCO 2015). According to the guidelines, the status of a World Heritage property becomes threatened due to improper management systems and the insertion of projects attempting the historical integrity of the property.

3 THE CASE OF CUENCA

3.1 *Planning and heritage*

3.1.1 *National level*
In Ecuador, during the past decade and after the government change in 2007, planning has played a significant role. Amendments during this presidential term (2007–2017) include the reform to the National Constitution in 2008, where, for the first time, a national planning system is established as the instrument for social change (ANC 2008; Sandoval 2015). Education and health played a leading role, although its effectiveness in the development process is firmly under discussion (Aldulaimi & Mora 2017; Gudynas 2013).

The 2008 National Constitution also instituted two critical changes for embedding spatial planning as state policy: 1) the notions of a Good Living Regime (Régimen del Buen Vivir) and 2) the concept of decentralisation. The first consists of an unprecedented contribution in favour of promoting the development of the country in respect of traditions and nature, while the second ends a longstanding tradition on centralisation of power and encourages the autonomy of sub-national governments and the active participation of civil society.

Once established the 2008 National Constitution, the Good Living Plan time-bound from 2009 to 2013 was approved and continued in force after the presidential re-election for the 2013–2017 term. This plan is linked to the National Decentralised Participative Planning System (SNDPP by its initials in Spanish), in charge of organising the development of an agenda in all governmental levels: national, regional, provincial, cantonal and parish (referred to the administrative-territorial organisation) level (Vivanco Cruz 2016).

Figure 1 shows a summary of how legal framework, territory and instruments for planning and development in Ecuador are related after the reform to the

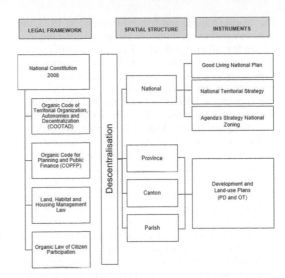

Figure 1. A scheme on the spatial planning instruments for development planning in Ecuador. Source: author, developed from Sandoval 2015, Vivanco Cruz 2016

National Constitution in 2008. The scheme depicts the 2008 National Constitution as the rector document, after which the creation of the Organic Code of Territorial Organization, Autonomies and Decentralization (COOTAD) was established to regulate the competences at different territorial levels (from national to parish). These levels correspond to the transversal idea of decentralisation inserted since 2008.

The instruments for implementing the legal framework relays on the Good Living National Plan, and due to the decentralisation process, local plans belong to its spatial territorial competence. Previous provincial, cantonal (municipalities) and parish bodies for management are replaced by Autonomous Decentralised Governments (GADs) that spatially correspond to the same territorial levels for planning: provincial, canton and parish level.

Historically, in Ecuador, fertile public policies for urban development and spatial planning traces back to the mid-twentieth century. The regional approach to planning characterises this period. Later, in the 1970s, urban centres became crucial for spatial planning in the country and set the stage for strategies to integrate urban and rural development during the 1980s. The heyday of the urban organisation was undertaken for the next decade until the national crisis of the 1990s. A turning point occurred at the end of this decade into an environment of political instability. The new millennium entailed urban studies arouse less apparent interest; while major cities were beyond their spatial capacity limit (Bermúdez et al. 2016). In this context, the reformed 2008 National Constitution mobilises the demands for control in spatial planning, moving from traditional regional planning towards the multi-level territorial approach.

In the heritage field, since Ecuador joined the 1972 UNESCO Convention concerning the protection of the World Cultural and Natural Heritage in 1975, the government has committed to comply with the requirements for cultural and natural heritage identification, protection, and management. Effective nomination processes have made way for the incorporation of five properties on the WHL, including the city of Quito and the Galapagos Islands in 1978, the Sangay National Park in 1983, the Historical Centre of Cuenca in 1999, and the Qhapaq Ñan, Andean Road System in 2014.

However, cultural heritage concern in Ecuador was registered previously around the end of the 19th and beginning of the 20th century. The creation of the History Academy, for example, dates from 1916, while the first heritage regulation from 1927 (Pedregal et al. 2014). This first regulation document inspired the official Heritage Law of 1945 and the creation of a national monitoring association, Casa de la Cultura (The House for Culture). The 1945 law was replaced by the 1978 Cultural Heritage Law, motivated by the first two inscriptions of Quito and the Galápagos Islands on the WHL. Additionally, this year, the National Institution of Cultural Heritage (INPC by its initials in Spanish) was created. Both legal associations and the regulatory documents are still in force, with a codification to the Cultural Heritage Law made in 2004.

The governmental shift in 2007 included explicitly in the reform to the National Constitution of 2008, the protection of the natural and cultural heritage as one of the primary duties of the State (ANC 2008). The COOTAD, as the legal framework in charge of the implementation of the National Constitution, allocates heritage management to the corresponding GADs at the different territorial levels for planning: provincial, canton and parish level.

The updated legal framework in 2008 propitiated the proposal of an Organic Culture Law Project in 2009 to replace the 1978 Cultural Heritage Law. The approval of this law entered in force in January 2017 (El Universo, n.d.).

Besides the institutionalised practice for spatial planning and heritage In Ecuador, citizen participation began to be considered within the 1998 Constitution. However, it was not until the 2008 Constitution that it was consolidated to promote mechanisms of participation in planning processes, projects execution and assessment, public decision-making, and social control. In practice, citizens have actively been part of elections processes. However, they have not extensively benefited from the rest of the participatory instruments as popular regulatory initiatives, public oversight, amongst others. According to Calderón (2015), this fact is due to the lack of massive education on participation rights.

3.1.2 Local level (Cuenca Canton)

Cuenca is a canton located on the southern highlands of Ecuador. It territorially is subdivided into 15 urban and 21 rural parishes within an area of 375.443,11Ha with a population exceeding the half a million of inhabitants (Ilustre Municipalidad de Cuenca & Universidad del Azuay 2011). According to Vivanco

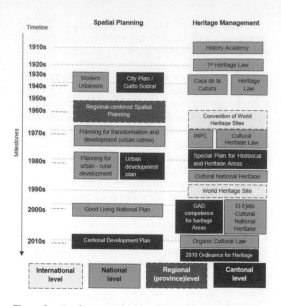

Figure 2. A scheme on the historical milestones for spatial planning and heritage management in Ecuador in the different governance levels. Source: author

(2016), Cuenca is amongst the first cantons to initiate the commended planning process referred to as the reformed 2008 National Constitution. The Cantonal Development Plan was completed in 2010 with 2030 as the horizon year. The plan is based on six strategic objectives: environmental; socio-cultural; financial; human settlements; mobility, energy and connectivity, and political, participative, and institutional systems (Ilustre Municipalidad de Cuenca 2016).

Historically, the first City Plan was developed in 1947 by the Uruguayan architect Gilberto Gatto Sobral who brought the modern era to the city. This plan launched a new urban structure in the area known as El Ejido located just on the border of the historic city centre. The former grid pattern used in the historic area is not used anymore in the plan of Sobral. Instead, the modern conception with the construction of a boulevard and the foresight of multiple public spaces was developed. Although not all the components of the plan were completed, the modern era concluded a longstanding tradition of urban settlement in Cuenca since its foundation in 1557. Although this City Plan did not overcome the urban layout of the historic areas, the modern spirit that brought with it meant the replacement of historic buildings at the core of the historic city.

The heritage values of this historical city were, for the first time, recognised through an initiative proposed by local architects. They developed the first inventory of valuable buildings in 1975 aimed at stopping the destruction of local architecture as the result of the arrival of modernity in the city (Jaramillo & Astudillo 2008).

Later, in 1982, a new Urban Development Plan (Canton-level) was adopted. Heritage management was considered as one of the key topics on this plan.

This fact facilitated the development of the first Special Plan for the Historical City Centre. Also, an updating of the inventory of heritage buildings in the historical areas and its declaration as National Cultural Heritage took place in the same year. This plan was replaced by the Cantonal Development Plan in force since 2010. In this plan, a link between planning and participatory processes is envisaged.

Regarding the local management, an ordinance for the control and administration of the historical areas was launched in 1983 (Lloret 2015) and replaced by the 2010 latest official regulatory ordinance document. The Historic Centre of the city of Cuenca is part of the WHL since 1999. This year an update of the inventory as well as a proposal for a new Special Plan were developed. The latter was not approved, and up to date, the 1982 Special Plan is still in force.

In the local experience, it is clear the mislinkage of planning and heritage conservation processes, the same that undertake actions according to divided perspectives. On the one hand, the reformed structure of the planning instruments at the national level meant an appropriation of local GADs of regulatory plans but still land-use based plans in which other topics are not included. On the other hand, in the heritage area, the local GAD does not envisage the planning of the Historic City Centre with its interlinkages with the canton and province. The Special Plan focuses only on the areas nominated as part of the territory selected for the WHL. A slight improvement in this aspect was evident in 2008 when, for the first time, an inventory of modern architecture in the area of El Ejido was carried out, and a minor sector attached to the Historic City Centre was declared as National Heritage. Figure 2 gives an overview historical milestones for spatial planning and heritage management in Ecuador in the different governance levels.

3.2 Actors in the governance levels in Cuenca

Based on the previous section, the different instances and actors involved in both fields of planning and heritage are defined in this heading. The competence of the spatial structure of Cuenca Canton is of the GAD, formerly known as the Municipality of Cuenca. Within the GAD, the dependencies that have a direct influence on spatial planning and heritage management are the Cantonal Council and the Directions of Planning, and the Historical and Heritage areas. The Planning one aims to consolidate and balance territorial development in the entire canton. The Direction of Historical and Heritage Areas envisions a sustainable process of planning and management within the residential, urban-architectural-landscape, equipment, and economic areas. The first is in charge of the Cantonal Development Plan, and the latter develops the Special Plan for the Historic Centre.

Besides the Direction of Historical and Heritage Areas, the Commission of the Historical Centre (CHC) was created in 1982 during the declaration of Cuenca

as National Heritage to safeguard the city's heritage. The INPC, which in 1982 had a direct responsibility on heritage management at the national level, delegated the protection of the Historical Centre to the Municipality of Cuenca. In turn, the CHC was formed (ICOMOS 1999). The CHC conceived as an inter-institutional entity in charge of the execution of the Special Plan, was designed in the framework of the 1982 Urban Development Plan, controlling the territory fixed for the Heritage Areas and authorising all interventions regarding heritage buildings. The CHC is an advisory body that consists of delegates from the following instances: Cantonal Council (2 councillors), INPC, Universities, citizen's representative, and a canton parish council delegate. While the Direction of Historical and Heritage Areas acts as the secretary of the CHC, any municipal worker or particular person can participate upon request. It is, therefore, this Commission, the space for consensus and participation of representative actors coming from organised and non-organised social actors. However, since both the Direction of Historical and Heritage Areas and the CHC are advisory bodies, the actual decision on heritage issues takes place only at the Canton level within the correspondent Council. Therefore, the decision-making process falls into political hands.

Once the GADs was awarded the competence of heritage management after the Constitutional reforms in 2008, the INPC functions as an advisory body with a regional competence -Region 6- (3 provinces in the southern area of Ecuador). In the same regional territory, Planning Zone 6 has competence in spatial planning.

Despite the levels of governance in both fields overlap (Table 1), they do not have a meeting or encounter point. Only for the 2010 ordinance, a transitory (three-months) advisory body was created, the Commission of Heritage Management, consisting of the Direction of Historical and Heritage Areas, the General Secretariat for Planning, and the Director for Urban Control.

This perspective on the institutionalised actors suggests that any particular citizen or representative group could be involved, for example, in the sessions of the CHC. However, even when individuals participate in these meeting places, the CHC remains as an advisory body, pending a final political decision.

4 HERITAGE INTEGRATION IN PLANNING, A CRITICAL ASSESSMENT

In Cuenca, both planning and heritage conservation evolved in stages that can be traced to understand turning points and the nowadays relation between planning and heritage through the actors involved at the different governance levels. Table 1 summarises crucial stages traced from previous sections, the governance levels, and actors involved. The first stage consists of the first city plan in 1947, which meant the replacement

Table 1. Critical stages of spatial planning and heritage management according to the different governance levels and actors involved in Cuenca. Source: author

Stage	Governance level	Actors involved in planning	Actors involved in heritage
First City Plan in 1947	International	International urban planners	
Planning and heritage concern in 1982	National		INPC
	Cantonal	Local consultants	Local consultants
		The Direction of Planning	The Direction of Historical and Heritage areas
			CHC
World Heritage Declaration in 1999	International		World Heritage Committee
	National		Minister of Education
	Cantonal		Local consultants
			The Direction of Historical and Heritage areas
State governance changes after the 2008 National Constitution	National	National Secretariat of Planning and Development	Minister of Culture and Heritage
		Planning Zone 6	INPC – Regional 6
	Local GAD (replaces the Municipality)	The Direction of Planning	The Direction of Historical and Heritage areas
			CHC
		Cantonal Council	Cantonal Council

of historic structures with the arrival of the modern era. This phase leads to the 1982 stage, where the emerging local and international debate on heritage and new planning for the city developed conciliating common concerns in both fields. In 1999, the international recognition led by local actors made possible the inclusion of Cuenca on the WHL. This fact propitiated the definition of the competencies for heritage management but did not upgrade the 1982 Special Plan for the Historical City Centre. The restructuring of roles and competencies in each governance level characterises the stage after the reforms of the 2008 National Constitution.

As suggested in the Operational Guidelines for world heritage sites, in Ecuador, different institutions have been created to safeguard heritage through the National Legislation. Nevertheless, the suggestion of integrating heritage protection into comprehensive planning programs is a pending task in the case of Cuenca. The integration involves the collection of all aspects of cultural heritage that can facilitate a better quality of life for citizens, meaning a direct impact on spatial planning. For this reason, it becomes crucial to integrate the goals of heritage conservation to the ones referred to spatial planning in the city. This crucial step follows a more explicit linkage of the social actors involved.

The lack of interplay amongst institutional actors can result, for example, in a duplicate effort for data collection used for different purposes. The INPC, the Planning Zone 6, the Direction of Planning, and the Direction of Historical and Heritage areas develop geographic information systems (GIS) in regional, cantonal, and historic areas. All of them are valuable data that could be put together and jointly used at all levels. An initiative in Chile shows the integration of heritage within the National GIS inferring interinstitutional agreements in an unprecedented effort to strengthen multi-disciplines capacities. The results show partial success since it still depends on peoples' voluntarily (Ladrón de Guevara González 2004).

In contrast to the aim of multi-level governance, the transformation of the institutional arenas in the case of Cuenca did not address an improved dynamic actors' interplay but instead has propitiated a political step in front of the opportunity of a consensus decision coming from the CHC, academic, and private initiatives. From the current exploration, it is apparent the mislinkage between spatial planning and heritage conservation practices at the different governance levels. However, it is also seen as an opportunity to overlap on the territorial structures in which both fields work.

Actions for the inclusion of civil society as an essential group in decision-making integrating spatial planning and heritage management have emerged in the last years in academia ('Cuenca, Ecuador – GO-HUL', n.d.) or sponsored by international entities ('CUENCA RED', n.d.). In the first case, one of the initiatives at the academic level is the World Heritage City (CPM by its initials in Spanish) research project at the University in Cuenca that since 2007 has promoted participatory activities in heritage topics in conjunction with the INPC and Direction of Historical and Heritage Areas. Within this project, a plan to implement the UNESCO Recommendation on Historic Urban Landscapes started in 2014. This Recommendation gathers the demanded integration of spatial planning and heritage. It offers a six-step action that includes - amongst others- the mapping of heritage resources, the reach of consensus and the integration of the above into an urban development framework. The implementation of the Recommendation in Cuenca led to the development of the first two steps via a participatory methodology known as sociopraxis (Rey-Pérez & Tenze 2018). This method includes active listening, feedback, reflection and joint action, framed in an integral, participatory, synergic and continuous process (Villasante 2010). Workshops with the community in 14 neighbourhoods, a photography and painting competition, as well as gathering event involved civil society coming from children, youth, public and private institutions. The results of these activities include a collective recognition of heritage outside of the historical city centre with an emphasis in public areas. This fact means a social reassessment of the city, where not only the inventoried buildings are recognised as heritage, but also areas outside the city centre, both tangible and intangible manifestations.

Through this initiative, the potential of participatory methods to integrate spatial planning and heritage is revealed. Thus, heritage management exclusively through a Special Plan for the Historical City Centre might not be enough when fundamental aspects of spatial planning go beyond its borders.

5 CONCLUSIONS

Strategic planning shows efficacy when properly context-specific implemented. In Latin America, the initial steps of using Strategic Planning embraces the institutionalisation in many cases of adopted external models. International guidelines as the 2011 UNESCO Recommendation for heritage management are also widely spread among State Parties highlighting the importance of adopting context-led actions. Context is a social construction, and therefore, the success of strategic planning and heritage management can be determined by a concerted dialogue with key actors since they will play an essential role in the implementation process.

Institutionalisation has been the leading force for heritage conservation and planning regulations in Cuenca. Although the levels of governance in Cuenca represent a political-led decision-making process, the role of the CHC is the potential space to re-open the broader participation of citizens since its conception relies on a multi-disciplinary and inclusive approach. However, before only relying on the effectiveness of this advisory body, it is needed to evaluate its impact after 38 years of existence. It is desirable that, at the CHC, citizen participation re-emerges in a leading role with, for example, a regular citizen observatory present in all sessions. Academic and private initiatives prove to be spaces to bridge community and social organised and non-organised institutions.

The spaces created by academia, as well as international cooperation activities, stimulate participation projects. Such projects should be encouraged to explore good practices and to exploit recommendations adapted to local contexts. Public-private cooperation should tackle problems at the historical areas promoting social involvement not only via participation but also through capacity building. These initiatives need support from the GAD and not only be narrowed towards areas in the historic centre but also consider the city as a whole.

The exploration carried out in this article demonstrates that multi-level governance is already present in Cuenca in similar territorial structure for both spatial planning and heritage management; however, further steps to implement social participation need to be tackled. So far, institutional structures have failed to create meeting spaces with actual impact on decision-making. Meanwhile, bottom-up initiatives lack the power to influence at the level of planning in the city.

REFERENCES

Albrechts, L., Healey, P., & Kunzmann, K. R. 2003. Strategic spatial planning and regional governance in Europe. *Journal of the American Planning Association*, 69(2), 113–129.

Albrechts, L. 2004. Strategic (spatial) planning reexamined. *Environment and Planning B: Planning and Design*, 31(5), 743–758.

Aldulaimi, S., & Mora, F. E. 2017. A Primary Care System to Improve Health Care Efficiency: Lessons from Ecuador. The Journal of the American Board of Family Medicine, 30(3), 380–383.

ANC. Constitución de la República del Ecuador Act, Pub. L. No. 449, §s 2 2008.

Andre, I., Abreu, A., & Carmo, A. 2013. 18. Social innovation through the arts in rural areas: the case of Montemor-o-Novo. In The International Handbook on Social Innovation: Collective Action, Social Learning and Transdisciplinary Research (Frank Moulaert, Diana MacCallum, Abid Mehmood and Abdelillah Hamdouch, pp. 242–255). Edward Elgar Publishing Limited.

Ashworth, G. 1994. From history to heritage—from heritage to identity. Building a New Heritage: Tourism, Culture and Identity in the New Europe, 13–30.

Bermúdez, N., Cabrera, S., Carrión, A., Del Hierro, S., Echeverría, J., Godard, H., & Moscoso, R. 2016. La investigación urbana en Ecuador (1990–2015): cambios y continuidades. P. Metzger, J. Rebotier, J. Robert, P. Urquieta y P. Vega Centeno (edits.), *La cuestión urbana en la Región Andina. Miradas sobre la investigación y la formación*, 117–173.

Calderón, F. A. C. 2015. La participación ciudadana y el control social en Ecuador. *Uniandes Episteme*, 2(1), 047–065.

Carrión, F. 2014. La ciudad y su gobierno en América Latina. Procesos urbanos en acción, 45.

Cuenca, Ecuador – GO-HUL. (n.d.). Retrieved 14 June 2017, from https://go-hul.com/2016/11/08/cuenca-ecuador/

CUENCA RED. (n.d.). Retrieved 14 June 2017, from http://cuenca.red/about/

Degen, M., & García, M. 2012. The transformation of the 'Barcelona model': an analysis of culture, urban regeneration and governance. International Journal of Urban and Regional Research, 36(5), 1022–1038.

El Universo. (n.d.). *En el 2017 regiría Ley de Cultura.* Retrieved 14 June 2017, from http://www.eluniverso.

com/vida-estilo/2016/03/11/nota/5455839/2017-regiria-ley-cultura

Gudynas, E. 2013. El malestar moderno con el buen vivir: reacciones y resistencias frente a una alternativa al desarrollo (Análisis).

ICOMOS. 1999. WHC Nomination Documentation of the Historic Centre of Santa Ana de los Rios de Cuenca (No. WHC-99/CONF.209/11). Paris: UNESCO. Retrieved from http://whc.unesco.org/uploads/nominations/863.pdf

Ilustre Municipalidad de Cuenca. 2016. Plan de ordenamiento territorial. Retrieved 30 May 2017, from http://www.cuenca.gob.ec/?q=page_planordenamiento

Ilustre Municipalidad de Cuenca, & Universidad del Azuay. Formulación del Plan de Desarrollo y Ordenamiento Territorial 2011.

Jajamovich, G. 2013. Miradas sobre intercambios internacionales y circulación internacional de ideas y modelos urbanos. Andamios, 10(22), 91–111.

Jaramillo, D., & Astudillo, S. 2008. Análisis de los inventarios del patrimonio cultural edificado en la ciudad de Cuenca. In Universidad de Cuenca. Facultad de Arquitectura 50 Años de La Universidad de Cuenca, 222–225.

Ladrón de Guevara González, B. 2004. Patrimonio y territorio: huellas del aprendizaje en tres años del Área de Patrimonio del Sistema Nacional de Información Territorial (SNIT). Conserva: Revista Del Centro Nacional de Conservación y Restauración, (8), 71–86.

Lloret, G. 2015. Cuenca: Patrimonio Mundial a 15 años de su declaratoria. Estoa. Revista de La Facultad de Arquitectura y Urbanismo de La Universidad de Cuenca, (6), 81–87.

Metzger, J., & Schmitt, P. 2012. When soft spaces harden: the EU strategy for the Baltic Sea Region. Environment and Planning A, 44(2), 263–280.

Moulaert, F., Martinelli, F., González, S., & Swyngedouw, E. 2007. Introduction: Social innovation and governance in european cities urban development between path dependency and radical innovation. European Urban and Regional Studies, 14(3), 195–209.

Murzyn-Kupisz, M., & Działek, J. 2013. Cultural heritage in building and enhancing social capital. Journal of Cultural Heritage Management and Sustainable Development.

Pedregal, A. M. N., García, C. S., & Segarra, E. C. 2014. Conversatorio sobre interculturalidad y Desarrollo. Universidad Miguel Hernández.

Pradel Miquel, M., García Cabeza, M., & Eizaguirre Anglada, S. 2013. Theorising multi-level governance in social innovation dynamics. In The International Handbook on Social Innovation: Collective Action, Social Learning and Transdisciplinary Research (Frank Moulaert, Diana MacCallum, Abid Mehmood and Abdelillah Hamdouch, pp. 155–168). Edward Elgar Publishing Limited.

Rey Pérez, J., & Tenze, A. 2018. La participación ciudadana en la Gestión del Patrimonio Urbano de la ciudad de Cuenca (Ecuador). Estoa. Revista De La Facultad De Arquitectura Y Urbanismo De La Universidad De Cuenca, 7(14), 129–141.

Sandoval, M. F. L. 2015. El sistema de planificación y el ordenamiento territorial para Buen Vivir en el Ecuador. GEOUSP: Espaço e Tempo (Online), 19(2), 296–311.

Situación GAD Cuenca. (n.d.). Retrieved 18 October 2014, retrieved from http://www.cuenca.gov.ec/?q=page_situacion

Steinberg, F. 2005. Strategic urban planning in Latin America: Experiences of building and managing the future. Habitat International, 29(1), 69–93.

UN Habitat. 2016. Urbanisation and development: Emerging Futures. World Cities report. UNESCO. Paris

UNESCO 2015. Operational Guidelines for the Implementation of the Convention of Cultural and Natural Heritage. UNESCO. Paris.

Vandesande, A. 2017. Preventive Conservation Strategy for built heritage aimed at sustainable management and local development. (PhD dissertation). KU Leuven.

Vivanco Cruz, L. 2016. La participación ciudadana: Una visión sistémica en el marco del ordenamiento territorial. Cuenca-Ecuador: Universidad de Cuenca.

Villasante, T. R. 2010. Stories and approaches of a participatory methodological articulation. CIMAS Notebooks-International. Retrieved from http://goo.gl/wpmabJ

Zukin, S. 2012. The social production of urban cultural heritage: Identity and ecosystem on an Amsterdam shopping street. City, Culture and Society, 3(4), 281–291.

The Future of the Past:
Paths towards Participatory Governance for Cultural Heritage – García et al (eds)
© 2021 Taylor & Francis Group, London, ISBN 978-1-032-02129-4

Designing transition spaces for sustainable futures: SARAS Transition Lab

C. Zurbriggen
University of the Republic, Uruguay SARAS Institute, Uruguay

S. Juri
Carnegie Mellon University, USA SARAS Institute, Uruguay

ABSTRACT: The various socio-ecological crises that characterize these transitional times demand new ways of understanding, thinking and acting. In this paper, we put forward a program for the creation of a transition lab in the context of the South American Institute for Resilience and Sustainability Studies. By drawing inspiration and integrating the approaches of Transition Design, Resilience Thinking and Policy Design, we offer a model of an experimental space for transdisciplinary and trans-sector collaboration with multiple actors from academia, public and private sectors, as well as civil society.

1   INTRODUCTION

In the last decade, social and public innovation labs have become increasingly visible instruments in innovation, collaborative work, and experimentation in Latin America (Acevedo & Dassen 2016; Ferreira & Botero 2020; Zurbriggen & Lago 2014). Since labs have gained interest as a way to promote dynamic responses and implementation of solutions, improve public engagement, and manage limited public funds more efficiently, their popularity has helped to bring attention to new approaches of scaling social innovation initiatives as well as promote experimentation in public policy. While some relate to the transformation-lab (T-Lab) approach as spaces that develop propositions (systems, products, services, policies, processes) adopting a methodology that enables collective exploration, creation and testing of diverse ideas (Bulkeley et al. 2016; Binder & Brandt 2008; Pereira et al. 2018), many initiatives found in the Latin American context tend to focus on promoting an entrepreneurial spirit and technological innovation or change. In general, there has been limited engagement with socio-environmental issues more explicitly (as we find in socio-ecological laboratories and experiments) with the exception of projects promoted by STEPS LatinAmerica[1], UNAM's Laboratorio de Ciencias de la Sostenibilidad[2] (Mexico), Proyecto Regional Transformación Socioecológica[3] and the series of SEGIB's Laboratorios de Innovación Ciudadana[4] (Costa Rica, Mexico, Argentina, Colombia, Brazil), to name a few. However, in a region with increased inequity and several environmental, economic and health challenges linked to extractivist models of production and development, adopting lenses that acknowledge the intricate relations between these issues becomes imperative to secure a sustainable and resilient future.

In the context of current socio-ecological crises, emerging perspectives often based on long traditions of thought and activism are attempting new ways of knowledge integration and action that adopt varying degrees of engagement with tacit and 'expert' knowledge. For over ten years, the South American Institute for Resilience and Sustainability Studies (SARAS Institute) has been acting as an interdisciplinary research institute with a regional scope and international community of practice seeking to generate critical insights allowing South America to build sustainable futures[5]. With an initial theoretical and applied direction anchored on socio-ecological systems transformation (Folke et al. 2010) and Resilience Thinking (Folke et al. 2016; Olsson et al. 2014), the integration of knowledges across fields and domains ranging from the social and natural sciences to the arts has been one of its pillars, displaying an open attitude to constantly incorporate new and diverse approaches. In this context, an opportunity has emerged to integrate and engage with transition theory (Geels 2005; Loorbach et al. 2017), and in particular, Transition Design (Irwin 2015) and Policy Design (Peters 2018), areas that offer various insights and tools when exploring the inherently material, social and political aspects existent in all socio-technical-ecological systems (Ahlborg et al. 2019) transition processes.

Given an opportunity to bridge and complement these different approaches, a proposal for the development of a transition laboratory at SARAS Institute stems from an interest in fundamentally rebuilding the science/state/society-mark nexus in an innovative and action-oriented way with different forms of citizen mobilization, new collaborative arenas and new forms of knowledge organization (co-produced, transdisciplinary) and power distribution. The complexity of our current problems challenges us to think about new ontological-epistemological arrangements (Leff 2009) which require a complete transformation of the matrix of understanding including the power relations that allow them to develop knowledge from social practices (Chomsky et al. 2006). Therefore, this lab's approach seeks to develop a reflexive culture that goes beyond the positivist vision (rationality, reductionism, predictability, determinism) to reflect 'in' and 'on' practice. It proposes an open-ended inquiry that conceives research as a collaborative process of tackling issues based on deliberation, experimentation, learning and context specificity, in which actors are led to question and jointly reframe their values and understanding (Dewey 1927; Schön 1984). This implies a new political epistemology of social intervention based on a culture of experimental pragmatism that allows us to deal with complex problems with a systemic perspective to generate social transformations (Ansell & Geyer 2017; Dewey 1927; Schön 1984). However, the complex and interlinked nature of the serious social, economic, and environmental issues that human societies currently confront demands responses that fully and equally engage with the social and environmental domains, moving beyond modern dichotomies[6] and either/or problem framings.

Approaches such as Resilience Thinking (Olsson et al. 2014) Transition Design (Irwin 2015) and the emerging public policy propositions (Peters 2018) are trying to overcome dichotomies and strive for knowledge plurality and synthesis. While adopting systemic perspectives, they in turn address the articulation of the needs, visions and aspirations of the different human and non-human actors, structures and systems in an ongoing process of interaction and regeneration of agencies in permanent transformation. In this paper, we draw inspiration from these approaches and put forward our own program for the creation of a Transition Lab at SARAS Institute, conceived as an experimental space for transdisciplinary and trans-sector collaboration with multiple actors from academia, public and private sectors, as well as civil society. The structure of this article moves from introducing the main definitions and frameworks to exemplify how the lab is modelled. The first section introduces the main theoretical contributions upon which the lab is created, meant to be used as a compass to begin identifying the capacities needed to develop a more robust framework for enabling transitions. Inspired by these approaches, the second part presents the evolving model developed for SARAS Transition Lab, with a particular focus on socio-environmental issues. We conclude by sharing reflections on the challenges and opportunities presented by adopting this strategy and offer insights into future actions and pathways for development.

2 SARAS TRANSITION LAB EVOLVING MODEL

2.1 Main definitions and conceptual pillars

Our Transition Lab is conceived as a platform to promote critical reflection and collective learning over strategies to address complex and systemic challenges. As a 'living space', the lab seeks to identify and produce new ways of understanding and engaging with the world, enabling the possibility to question assumptions and values, and foster the imagination and materialization of alternative futures pathways. Emergent ideas and visions are thus transformed into public value for society. In this sense, this platform is understood as a requirement to facilitate or support ongoing or new transition processes. We have chosen the laboratory format to bring forth an open and experimental space devised to explicitly explore and facilitate transitions to sustainable futures through learning-focused actions where it is safe to ask and explore "what if" questions. We particularly adopt Transition Design (Irwin 2015), an emerging perspective seeking to facilitate societal transition processes by supporting or developing interventions to intentionally change values, technologies, social practices, and infrastructures while reshaping the interactions between socio-technical and socio-ecological systems (Ceschin & Gaziulusoy 2016) in the context of people's everyday lives.

In advancing a broad, fluid and plural evolving body of knowledge (Irwin et al. 2015), transition design proposes new ways of thinking, being and doing, while supporting and creating new opportunities by design. Its tools and practices amalgamate theory and mindsets that cross various fields and knowledge systems (from living systems and complexity theory to post-normal science, from transition theory and social practice theory to indigenous knowledge, among others), and promote collaborative spaces of practice, learning, engagement and experimentation. Its reflective and practical (Schön 1984; Steen 2013) approach to dealing with systemic issues offers a way to envision and truly enact new collective onto-epistemic ways of being and knowing through action, engaging fully with the concept of the pluriverse –a world where many worlds (worldviews) fit (Escobar 2018). By adopting large spatio-temporal scales, Transition Design seeks to foster initiatives that can conform ecologies of actions –synergies– to support or disrupt system configurations that can be more appropriate and desireable. Such practical outcomes which may include material and symbolic interventions (with design mediating all interactions between humans and the more-than-human world) open opportunities for the development of whole new narratives and lifestyles (Irwin et al. 2015; Irwin et al. 2020), and by doing so, challenge unsustainable values and paradigms (Du Plessis & Cole 2011) through a process of self and collective learning and transformation.

With an explicit orientation towards systems level change, the lab is informed by theory and tools from different systemic approaches such as Resilience Thinking (Olsson et al. 2014; Folke et al. 2010) with its focus on stability or transformation of the beneficial relations between ecosystems and society, and sustainability transitions theory (Geels 2005; Loorbach et al. 2017) which comes to complement the socially constructed nature of socio-technical system assemblages of norms, structures, technologies and dynamics. Policy Design (Peters 2018) is also especially adopted given the need to address the normative and political aspects that allow, mediate or prevent all transitions processes. This emerging approach stresses the need to experiment in policy design with a more open, humane, systemic, anticipatory and experimental approaches (Ackoff 1974; Checkland & Scholes 1990; Rein & Schön 1994). By articulating these approaches, the lab seeks to generate a space to address socio-environmental problems by re-politicizing the space, considering the relevance of the restructuring of governance, power relations and empowerment while incentivizing new political capacities for transformation. This stresses the need to incorporate the dynamic interactions that exist between actors, power relations, controversies, conflicts and disagreements that are a fundamental part of any transition process (Forsyth 2018).

2.2 *Knowledge synthesis and transdisciplinarity*

While modernity has been a process attempting dichotomous bifurcations: natural science has been separated from social science, science from politics, nature from culture and intuition from reasoning (Kahneman 2011; Latour 2005; Österblom et al. 2015; Scheffer et al. 2015), such bifurcations are never fully achieved (Latour 1990). Still, since the cultural frame of modernity is predominantly taken as a reality, such framing is one of the major impediments when addressing complex problems (Snow 1959/1978). Transcending this barrier demands making explicit efforts to move towards transdisciplinarity and achieving knowledge synthesis. The Transition Lab is thus proposed as an open, transdisciplinary platform that allows for collaboration, reflection and collective learning over strategies to address systemic problems. Given that knowledge, learning and social change are interwoven in very complex and non-evident forms that require a high degree of integration and experimentation, our proposal for the lab explicitly seeks to connect science, art, design and politics while engaging plural voices and worldviews –multiple knowledge systems and paradigms. This integrative approach stimulates the balance of dual thinking (Scheffer et al. 2015) by associating analytical reasoning processes with intuitive and synthetic thinking to address the total dynamics of the system (see Figure 1 below) and reflect the reality of any decision-making process or practice.

Furthermore, by engaging knowledge systems outside of academia, especially necessary when dealing with problems that people face in their daily lives and particular contexts, the process acknowledges that stakeholders are highly valuable for knowledge co-creation by bringing their expertise, experiences and types of wisdom, all of which are necessary to properly understand and frame the problem at hand.

In a process where the 'consilience' between creativity and reasoning (Wilson 1998) is fundamental

ANALYSIS	SYNTHESIS
Separation	*Unification*
Rational	Emotional
Logical	Intuitional/pragmatic
Deductive	Inductive
Solutions	Paradigms, platforms
Thinking it through	Think through doing
One discipline	Multiple disciplines, T shape
Causality	Impact, value, diffusion

Figure 1. Differences between analytic and synthetic thinking. Adapted from Bason (2010).

to transcend barriers and make decisions, artistic and *designerly* (Cross 2001) approaches offer the potential to foster intuitive, experiential and less inhibited ways to explore and represent systems dynamics and people's positions in these dynamics (Curtis 2009; Curtis et al. 2012; Wiek et al. 2014). These approaches are an integral part of the lab's conception for three reasons. Firstly, artistic approaches help to integrate and make explicit the emotional aspects of environmental governance (Curtis 2009; Curtis et al. 2012; Scheffer et al. 2015; Wiek et al. 2014). Secondly, they help materialize and apprehend imagined scenarios –visions (Candy & Kornet 2019)– and conceive the steps that might lead to them. Thirdly, creative approaches serve as bridges to make explicit and strengthen existing connections between people, and between people and natural elements (Inwood 2008; Kagan 2008; Selman et al. 2010, cited in Heras et al. 2016). In combination, these features encourage different learning approaches that are highly explorative and motivating (Flowers et al. 2015; McNaughton 2004, cited in Heras et al. 2016; Scheffer et al. 2015). Perhaps even more importantly, they allow us to engage in meaningful discussions about the values that guide governance processes and help substantiate them in terms of care and concern, as opposed to mere facts. Indeed, the Transition Lab fully assumes that knowledge is not value-neutral and explicitly addresses the issue of values. Analytical thinking comes to complement the more creative exploration of values by making explicit whose values are being put forward, and which actors benefit from certain scenarios and solutions (Ansell & Geyer 2017).

In processes leading to social transformations, the quest for novel methods that can support means of transformational learning and people's empowerment is one of the main challenges. Within the Transition Lab, art and design approaches are adopted to foster: (a) a 'consilience' between knowledge, values and perspectives in multi-stakeholder dialogues, linking diverse knowledge fields with personal experiences, emotions and ethical judgments; (b) communication, translation and understanding of complexity; (c) social reflexivity, public deliberation and understanding; (d) the development of socio-ecological identities and ecological consciousness; and (e) engagement and emotional commitment leading to action (Heras et al. 2016).

2.3 *A space for creativity and experimentation*

The lab's experimental maxim seeks to promote a reflexive culture exploring different strategies for the integration of science, policy, art and design in order to build transition capacities. This implies a new political epistemology of public intervention based on a culture of experimental pragmatism that allows us to deal with complex problems with a systemic perspective and generate social transformations (Ansell & Geyer 2016; Dewey 1927; Schön 1984). As such, it is a space for being, thinking and doing together through the convergence of actors, talents and skills, ideas, policies, research and visual outcomes. To do so, this space tries to convene natural and social scientists, artists, other relevant stakeholders, and decision-makers to frame socio-ecological challenges and develop a shared-learning agenda. By adopting a critical systems perspective of knowledge to address complex issues, this platform seeks to re-connect humans and nature, transforming new ideas into practical action for generating new societal values (Macintyre et al. 2019).

We see the lab as a networked space where 'space' takes different meanings. It may refer to a physical space (e.g. an office or multi-location space), a virtual space (e.g. a skype-call), a mental space (e.g. shared experiences, ideas, ideals), or any combination of these. In any case, interaction is the most important aspect to allow for tacit knowledge sharing and development, especially in face-to-face exchanges, a feature which is emphasized in our model. In order to enable this space to take shape, our Transition Lab model is based on 4 principles: systemic, social, critical and generative (see Figure 3). Firstly, it seeks to make sense of the relationships and entities existing behind a complex problem, to enhance and broaden the appreciation of its dynamics rather than aim for thorough, detailed knowledge. Secondly, it seeks to actively engage diverse and contrasting perspectives (not necessarily all) so that possibilities for reshaping and improving the problem-situation can be identified for ongoing resolution as opposed to finding unique solutions. This is enhanced by fostering a process of social imagination and the adoption or large temporal scales –futuring (Mulgan 2020). Critical thinking comes to aid in exploring and reconciling ethical issues, power relations and boundary issues that are inevitable in partial understandings of systems when different stakeholders are involved (Midgley 2000). Finally, its generative aspect supports the experimental character of the process, where provisional and adaptive propositions are prototyped, tested and reformulated as a way to promote emergence and constant learning. In essence, the intention here is to gently disrupt, unsettle, and thereby provoke new systems thinking and acting (Reynolds 2011).

Engaging contrasting perspectives mean addressing and embracing conflict as an inherent and necessary aspect of any transformational process, as it especially determines the potential pathways or courses of action adopted. We rely on the contributions of critical systems thinking (Ackoff 1974; Churchman 1979; Checkland 1981; Midgley 2000) whereby systems are seen as constructs that aid understanding as opposed to real-world (objective) entities. Any systems intervention demands, according to the theory of boundary critique, reflecting on and choosing between different boundaries of analysis that shape how problematic situations are defined and managed (Churchman 1979; Ulrich 1996; Midgley 2000). Boundaries indicate what information is considered relevant and what is considered superfluous and are the result of value judgements. The exploration and setting of boundaries -how far they can be pushed out to allow for more information-can, according to Ulrich (1983), be undertaken through dialogue between stakeholders to make them more robust as opposed to being imposed by planners or external researchers in the absence of meaningful community participation. However, Midgley (2000) raises a relevant question: what happens when there are conflicts between stakeholders making different value and boundary judgements?

If we consider two boundaries (Fig.2), one narrower than the other one, a marginal area emerges between the two where we would find people or issues that are of concern to those operating with the wider boundary problem or system but which are excluded from the concerns of those using the narrower boundary. Because each boundary involves a set of ethics (values in action), conflict emerges, and the marginal elements become its focus. In such situations, conflict can be either productive –leading to resolution, or unproductive. In many social situations, resolution does not happen, and the conflict is instead stabilised, which is the result of the imposition of either a 'sacred' or 'profane' status on marginal people or issues. Because this status can be institutionalised by social rituals, the dominating value judgements would stabilise the conflict and determine whether the marginal elements become excluded, ignored or derogated (sacred narrow boundary, profane wider one) or integrated (sacred wider boundary) allowing for the narrower boundary to be challenged and lose strength. Because different stakeholders and issues can be marginalised and made profane for various reasons, Midgley's theory of boundary critique suggests the need to normatively prescribe a course of action (not just a mere description) that includes a reflection on boundaries during any intervention so as to avoid taking assumptions for granted. However, because boundary judgements are made by human beings, they are closely tied to value judgements, meaning that this type of boundary critique will help us reflect on different understandings of interventions as well as its moral purposes.

2.4 *Transformative change, learning and values*

Creating the conditions for enabling transitions towards sustainable futures requires radical, systemic changes in values, beliefs and in patterns of social behavior (Westley & Laban 2015) which in turn require an ontological-epistemological revolution that can re-embed the social spheres within nature. To embrace this degree of paradigm change, one of the most important features influencing the Transition

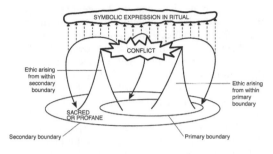

Figure 2. Marginalization Process (from Midgley 2015, p. 159, with permission).

Lab refers to transformative learning mechanisms (Mezirow 2000) where the examination of underlying assumptions leads to change in attitudes, behavior, and social norms. Transformative learning is a process in which participants with diverse *knowledges* are encouraged to co-create new strategies through experimentation, creativity, openness and trust. An environment that is conducive to sharing and team-building is key when enhancing participant's synergies for sharing information, skills and ideas. Furthermore, a learning space is one that combines adaptive learning: the capability to interpret, react, adapt to or influence our environment as necessary for survival; with generative learning which enhances the capacity to create (Senge 1990). A generative space is not a linear 'A to B' process but rather an emergent[7] one, empowering people to anticipate and create strategies in response to emerging issues and discontinuities.

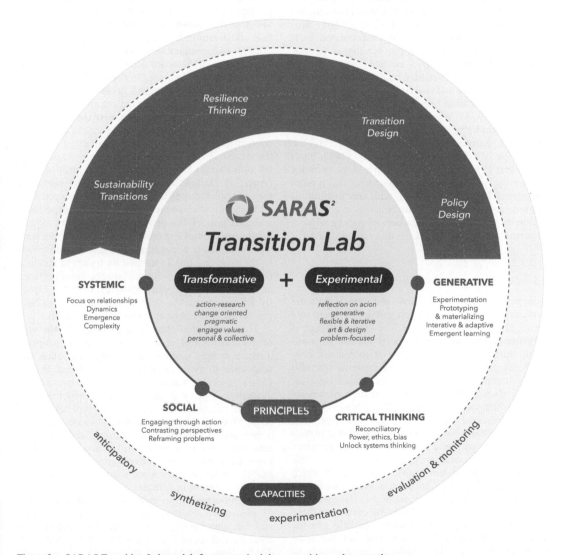

Figure 3. SARAS Transition Lab model: features, principles, capacities and approaches.

In complex processes, decisions must be taken with limited knowledge of the process's dynamics. Decision makers need to identify what critical knowledge is missing, and deal with uncertainties before committing time and resources to a particular action plan to develop a longer-term, systemic view. In such situations, the most effective strategy is to plan for learning as opposed to implementation. In this sense, a lab is often not about problem-solving and rather requires a more creative and values-based approach. This stance infers an openness to allow new strategies to come forth without shaping them according to our own beliefs or biases.

Following this logic, the Transition Lab is conceived as a space where actors would integrate values of solidarity, reciprocity and sustainability as a means to reach a moral-ecological well-being[8]. With a focus on deliberation, experimentation and context specificity, actors are encouraged to question and jointly reframe their values (Dewey 1927; Schön 1984). This transformation would equate to what Leff (2009, p. 105) identifies as 'a change of skin', moving past outdated linear ideas of the *Homo Economicus* paradigms towards enhancing the *Homo Reciprocans* and *Homo Sociologicus* (Dash 2016, p. 20) where human behaviors in the face or complexity and uncertainty are not determined by the maximization principle of utility (Tversky & Kahneman 1974). While the hegemonic conception of the human being as a selfish creature (Homo Economicus) involves the destruction of nature and the promotion of social inequality, Polanyi (1957/2001) stated that it is possible to develop a more complex relationship with society and the economy that is more altruistic (Homo Reciprocans), with new institutions to guide and encourage human behavior in a way that maintains the life of all members of society without undermining the Earth systems that support life.

2.5   *Developing capacities for transitions*

Adopting this type of systemic and transformational approach requires the adoption and development of a series of skills and capacities. The first challenge is to generate **the capacity for anticipation in managing uncertainty**. Uncertainty is rooted not only in the unpredictability of natural systems but also in the imperfect knowledge about human behavior as well as inherent variability and unpredictability of such behavior. Moreover, when multiple stakeholders are involved in co-creation processes, each comes with their own belief systems, views, preferences and interests, and thus their own interpretations of the same information. This gives rise to a new type of uncertainty –ambiguity, 'a situation in which a decision-maker does not have a unique and complete understanding to be managed' (Brugnach et al. 2009; Jensen & Wu 2017). To overcome these difficulties, anticipatory thinking can help provide a reflective dialogue from an intellectual and emotional perspective, which includes the discovery of the different interpretative frameworks (framings) and the worldviews underlying the value system that determine our actions. This further helps mobilize people to collectively identify and transform constructed visions into action.

The second challenge is to **generate capacity to synthesize knowledge** in a transdisciplinary way. These practices involve continuous interaction between actors from different social subsystems (research, politics, civil society, private sector) to link different perspectives and types of knowledge (scientific and experiential), in order to achieve a deeper understanding of the problem in the real life and generate a compass for better decision making (Pohl & Hirsch Hadorn 2008). The learning process involves the exploration and integration of useful knowledge, either tacit or codified, for a deeper understanding of a problem, better decision-making and, therefore, transformation (Westberg & Polk 2016). The most relevant conceptual base for transdisciplinarity is the systemic vision of the problem as a process of social construction and learning in action (Pohl & Hirsch Hadorn 2008) as two inseparable and simultaneous acts (Westberg & Polk 2016) which necessarily relate and interconnect facts, judgments, visions, values, interests, epistemologies, time scales, geographic scales and world views, all of which are not exempt from conflict.

The third challenge **is to generate the capacity to experiment,** to develop tangible spaces in the current context that allow change. Experimentation requires the development and use of a range of experimental tools that go beyond randomized controlled trials (Ansell & Bartenberger 2016) and adopt participatory and holistic approaches. A generative experiment addresses a particular problem rooted in the experience and situation of the people conducting the experiment (experiential and problem-oriented). There is no a priori certainty about singular and correct solutions to the problem, this is learned by trying to tackle it (learning by doing). 'Solutions' are continually refined as they are implemented (iterative) while the capacity for transformative implementation is being built (transformative) (Ansell & Bartenberger 2016).

The fourth **challenge is to innovate in the way of evaluating and monitoring processes**, adopting a new paradigm oriented to learning, innovation and adaptation in dynamic and complex systems. In processes involving various actors and interactions, it is unclear how or if an intervention will lead to a specific outcome. The inherent unpredictability of change requires integrative and adaptive management, a process of probing and learning, and recurring reflection on emerging patterns (Patton 2010). Patton specifically defines development assessment as one that informs and supports innovative and adaptive development in complex dynamic environments. This approach seeks to guide the collaborative action of innovative initiatives that face great uncertainty and that are characterized by their experimental, co-creation and social learning nature (Arkesteijn et al. 2015). Here, the unit of analysis for change, and therefore for evaluation, is no longer the project or program (as in conventional models) but the system. Because transitions unfold across long horizons of time and its elements are in constant flux, an iterative, adaptive and continuous learning attitude is required as a way to observe and assess change qualitatively, and thus be able to continually course correct based on the values, motivations and preferred future visions or transition pathways that each group normatively co-develops and recursively analyze. Because of this, the evaluation of the learning process developed by these multi-actor groups and networks is key (Zurbriggen & Lago 2020). To do so, we adopt the Reflexive Monitoring in Action (Mierlo et al. 2010), a methodology used to encourage learning and institutional change, especially adopted in reflexive governance processes dealing with complex problems. Because transition processes include stakeholders with different knowledge and understandings of issues, collaborative learning offers a space for the negotiation of meaning (by understanding contributions, accepting disagreement or delaying opinionated responses among other strategies) and the possibility to identify a 'common ground' where major differences or misunderstandings may be overcome (Van Mierlo & Beers 2020).

3 FINAL REFLECTIONS

By adopting the broad and evolving Transition Design approach as the compass that guides an integrative, plural and action-oriented engagement with systems level change, the model proposed here has attempted to outline the rationale, theoretical underpinnings, principles and mindsets that allow for the development of a laboratory platform to facilitate sustainability (socio-environmental) transitions in a Latin American context. Contextualizing this transdisciplinary space at SARAS Institute means adopting a clear direction that focuses on the complex interrelations that exist between social and environmental issues in a region where the development pathways adopted (at the individual and societal levels) have the potential to make or mar opportunities for healthy, just and desireable futures. Achieving moral-ecological well-being demands new approaches that integrate knowledge and action, challenging taken-for-granted matrixes of understanding, power relations and decision-making processes. It further involves a creative epistemological challenge of inductively understanding social phenomena (action-research) in specific socio-cultural and historical-temporal contexts which can be achieved through transdisciplinarity using qualitative, constructivist, and naturalistic approaches, as opposed to cartesian dichotomy and logical positivism (Dash 2016). In addition, a collective and normative process engaging plural and contrasting stakeholders demands addressing the politics of change, with the arising of different forms of conflicts as problem understanding, framing and reframing occurs. The narratives that are upholded,

problematized or reshaped throughout this process would further determine the type of framing that is possible, as well as the type of interventions or policy instruments that may be promoted or adopted. To achieve this type of space, attitudes and capacities, the lab proposes to develop a reflexive culture of open-ended inquiry based on deliberation, experimentation, learning and context specificity allows us to deal with complex problems with a systemic perspective to generate social transformations (Ansell & Geyer 2017; Dewey 1927; Schön 1984).

The two dimensions highlighted here signify an expansion in relation to the approaches already offered by Transition Design and Resilience Thinking, and therefore bridge theories and practices that may develop new insights to strengthen the political nature and agencies found in the type of socio-ecological transition that is sought here. Following the model with a truly experimental and fluid approach demands the generation of new collective spaces for social learning, imagination and action. Therefore, the first spaces that the lab has proposed relate to two initial lines of inquiry (sustainable food systems and integrative water management) and the articulation of two courses in collaboration with the University of the Republic of Uruguay and SARAS Institute. These courses will seek to materialize and develop the capacities and actions needed for the lab to continue to grow, evolve and transform. A broad understanding of space enables expansive fluidity, continuous emergence and evolution through regional and international networking, exchange and collaboration. This way, SARAS Transition Lab emerges as an open contextualized yet international platform to amalgamate views, experiences and connections, amplifying collective agency and wisdom, while enacting new ways of thinking and being in the world.

4 NOTES

[1] See STEPS (Social, Technological and Environmental Pathways to Sustainability) Centre Latin America chapter at https://steps-centre.org/global/steps-america-latina/

[2] See UNAM's Laboratorio Nacional de Ciencias de la Sostenibilidad Lab at http://lancis.ecologia.unam.mx/

[3] See Friedrich-Ebert-Stiftung (FES) Proyecto Regional Transformación Social-Ecológica at https://www.fes-transformacion.org/

[4] See Secretaría General Iberoamericana (SEGIB) projects at https://www.innovacionciudadana.org/

[5] See SARAS Institute at http://saras-institute.org

[6] As Bruno Latour puts it, modernity is a process that attempts at dichotomous bifurcations: natural science has been separated from social science, science from politics, nature from culture and intuition from reasoning (Latour 2005, Kahneman 2011, Scheffer et al. 2015, Osterblom et al. 2015).

[7]Emergence can be defined as "the arising of novel and coherent structures, patterns, and properties during the process of self-organization in complex systems"(Goldstein 2004). The evolution of the space is also influenced by the mechanisms they use to define their strategies and to implement them. All organizations have two simultaneous strategies: deliberate and emerging. The Lab is dominated by the emergent strategy.

[8]The Homo Reciprocans and Sociologicus are not the Homo Economicus of previous paradigms. These actors integrate values of solidarity, reciprocity and sustainability as a means to reach a moral ecological well-being. See also Lima & Fazzi (2018).

REFERENCES

Acevedo, S. & Dassen, N. 2016. Innovando para una mejor gestión. La contribución de los laboratorios de innovación pública. *Bid*, 66

Ackoff, R.L., 1974. *Redesigning the future: a systems approach to societal problems*. New York: Wiley.

Ahlborg, H., Ruiz-Mercado, I., Molander, S., and Masera, O. 2019. Bringing Technology into Social-Ecological Systems Research—Motivations for a Socio-Technical-Ecological Systems Approach. *Sustainability*, 11 (7), 2009–2009.

Ansell, C. & Geyer, R. 2017. 'Pragmatic complexity'a new foundation for moving beyond 'evidence-based policy making'? *Policy Studies*, 38 (2), 149–167.

Ansell, C.K. & Bartenberger, M. 2016. Varieties of experimentalism. *Ecological Economics*, 130, 64–73.

Arkesteijn, M., Mierlo, B. van, & Leeuwis, C. 2015. The need for reflexive evaluation approaches in development cooperation. *Evaluation*, 21 (1), 99–115.

Bason, C. 2010. *Leading public sector innovation: Co-creating for a better society*. Policy press.

Binder, T. & Brandt, E. 2008. The Design:Lab as platform in participatory design research. *CoDesign*, 4 (2), 115–129.

Brugnach, M., Henriksen, H., Van Der Keur, P., & Mysiak, J. 2009. Uncertainty in adaptive water management. *Concepts and guidelines. Osnabrück: University of Osnabrück*.

Bulkeley, H., Coenen, L., Frantzeskaki, N., Hartmann, C., Kronsell, A., Mai, L., Marvin, S., McCormick, K., van Steenbergen, F., & Voytenko Palgan, Y. 2016. Urban living labs: governing urban sustainability transitions. *Current Opinion in Environmental Sustainability*, 22.

Ceschin, F. & Gaziulusoy, I. 2016. Evolution of design for sustainability: From product design to design for system innovations and transitions. *Design Studies*, 47, 118–163.

Checkland, P. 1981. *Systems thinking, systems practice*. Chichester: John Wiley & Sons.

Checkland, P. & Scholes, J. 1990. *Soft Systems Methodology in Action*. Chichester: Wiley.

Chomsky, N., Elders, F., & Foucault, M. 2006. *La naturaleza humana: justicia versus poder. Un debate*.

Churchman, C.W. 1979. *The systems approach and its enemies*. Basic Books.

Cross, N. 2001. Designerly Ways of Knowing: Design Discipline Versus Design Science. *Design Issues*, 17 (3), 49–55.

Curtis, D.J. 2009. Creating inspiration: The role of the arts in creating empathy for ecological restoration. *Ecological Management and Restoration*, 10 (3), 174–184.

Curtis, D.J., Reid, N., & Ballard, G. 2012. Communicating Ecology Through Art: What Scientists Think. *Ecology and Society*, 17 (2), art3.

Dash, A. 2016. An epistemological reflection on social and solidarity economy. Presented at the Forum for Social Economics, Taylor & Francis, 61–87.

Dash, A. 2019. 'Good Anthropocene': The Zeitgeist of the 21st Century. *In*: A.K. Nayak, ed. *Transition Strategies for Sustainable Community Systems: Design and Systems Perspectives*. Cham: Springer International Publishing, 17–29.

Dewey, J. 1927. The Public and Its Problems (New York: H. Holt and Company).

Escobar, A., 2018. *Designs for the pluriverse: Radical interdependence, autonomy, and the making of worlds*. Duke University Press.

Ferreira, M. & Botero, A. 2020. Experimental governance? The emergence of public sector innovation labs in Latin America. *Policy Design and Practice*, 0 (0), 1–13

Folke, C., Biggs, R., Norström, A.V., Reyers, B., & Rockström, J. 2016. Social-ecological resilience and biosphere-based sustainability science. *Ecology and Society*, 21 (3).

Folke, C., Carpenter, S.R., Walker, B., Scheffer, M., Chapin, T., & Rockström, J. 2010. Resilience Thinking: Integrating Resilience, Adaptability and Transformability. *Ecology and Society*, 15 (4), art20–art20.

Forsyth, T. 2018. Is resilience to climate change socially inclusive? Investigating theories of change processes in Myanmar. *World Development*, 111, 13–26.

Geels, F.W. 2005. The dynamics of transitions in sociotechnical systems: a multi-level analysis of the transition pathway from horse-drawn carriages to automobiles (1860–1930). *Technology analysis & strategic management*, 17 (4), 445–476.

Heras, M., David Tabara, J., & Meza, A. 2016. Performing biospheric futures with younger generations: A case in the MAB Reserve of La Sepultura, Mexico. *Ecology and Society*, 21 (2).

Irwin, T. 2015. Transition Design: A Proposal for a New Area of Design Practice, Study, and Research. *Design and Culture*, 7 (2), 229–246.

Irwin, T., Kossoff, G., & Tonkinwise, C. 2015. Transition Design: An Educational Framework for Advancing the Study and Design of Sustainable Transitions, paper for the 6th International Sustainability Transitions Conference. *University of Sussex, UK*.

Irwin, T., Tonkinwise, C., & Kossoff, G. 2020. Transition Design: The Importance of Everyday Life and Lifestyles as a Leverage Point for Sustainability Transitions. *Cuadernos del Centro de Estudios en Diseño y Comunicación. Ensayos*, (105), 67–94.

Jensen, O. & Wu, X. 2017. Embracing Uncertainty in Policy-Making: The Case of the Water Sector ScienceDirect Embracing Uncertainty in Policy-Making: The Case of the Water Sector. *Policy and Society*, 35 (2), 115–123.

Leff, E. 2009. Degrowth, or deconstruction of the economy: Towards a sustainable world. *Critical Currents Dag Hammarskjöld Foundation*, (6).

Lima, J.A. de, & Fazzi, R. de C. 2018. A subjetividade como reflexividade e pluralidade: notas sobre a centralidade do sujeito nos processos sociais. *Sociologias*, 20 (48), 246–270.

Loorbach, D., Frantzeskaki, N., & Avelino, F. 2017. Sustainability Transitions Research: Transforming Science and Practice for Societal Change. *Annual Review of Environment and Resources*, 42 (1), 599–626.

Macintyre, T., Monroy, T., Coral, D., Zethelius, M., Tassone, V., & Wals, A.E. 2019b. T-labs and climate change narratives: Co-researcher qualities in transgressive action–research. *Action Research*, 17 (1), 63–86.

Mezirow, J. 2000. *Learning as Transformation: Critical Perspectives on a Theory in Progress. The Jossey-Bass Higher and Adult Education Series*. ERIC.

Midgley, G. 2015. Systemic intervention. *In: The Sage handbook of action research*. SAGE, 157–166.

Midgley, G. 2000. Systemic Intervention. *In*: G. Midgley, ed. *Systemic Intervention: Philosophy, Methodology, and Practice*. Boston, MA: Springer US, 113–133.

Mulgan, G. 2020. The Imaginary Crisis (and how we might quicken social and public imagination).

Mierlo, B.C. van, Regeer, B., Amstel, M. van, Arkesteijn, M.C.M., Beekman, V., Bunders, J.F.G., Cock Buning, T. de, Elzen, B., Hoes, A.C., & Leeuwis, C. 2010. *Reflexive Monitoring in Action. A guide for monitoring system innovation projects*. Wageningen/Amsterdam: Communication and Innovation Studies, WUR; Athena Institute, VU, No. 9789085855996.

Olsson, P., Galaz, V., & Boonstra, W.J. 2014b. Sustainability transformations: a resilience perspective. *Ecology and Society*, 19 (4).

Österblom, H., Scheffer, M., Westley, F.R., van Esso, M.L., Miller, J., and Bascompte, J. 2015. A message from magic to science: Seeing how the brain can be tricked may strengthen our thinking. *Ecology and Society*, 20 (4), 2–5.

Patton, M.Q. 2010. *Developmental evaluation: Applying complexity concepts to enhance innovation and use*. Guilford press.

Pereira, L.M., Karpouzoglou, T., Frantzeskaki, N., & Olsson, P. 2018. Designing transformative spaces for sustainability in social-ecological systems. *Ecology and Society*, 23 (4).

Peters, B.G., 2018. *Policy problems and policy design*. Edward Elgar Publishing.

Pohl, C. & Hirsch Hadorn, G. 2008. Methodological challenges of transdisciplinary research. *Natures Sciences Sociétés*, 16 (2), 111–121.

Polanyi, K. 1957/2001. *The great transformation: the political and economic origins of our time*. Boston: Beacon Press.

Rein, M. & Schön, D. 1994. *Frame Reflection Toward the Resolution of Intractable Policy Controversies (New York: Basic)*. New York: Basic Books.

Reynolds, M. 2011. Critical thinking and systems thinking: towards a critical literacy for systems thinking in practice. *In*: C.P. Horvath and J.M. Forte, eds. *Critical Thinking*. New York, USA: Nova Science Publishers, 37–68.

Scheffer, M., Bascompte, J., Bjordam, T.K., Carpenter, S.R., Clarke, L.B., Folke, C., Marquet, P., Mazzeo, N., Meerhoff, M., Sala, O., & Westley, F.R. 2015. Dual thinking for scientists. *Ecology and Society*, 20 (2), 1–4.

Schön, D.A. 1984. *The reflective practitioner: How professionals think in action*. Basic books.

Senge, P.M. 1990. *The fifth discipline: the art and practice of the learning organization*. 1st ed. New York: Doubleday/Currency.

Snow, C.P., 1959/1978. *The two cultures*. New York: Cambridge University Press.

Steen, M. 2013. Co-design as a process of joint inquiry and imagination. *Design Issues*, 29 (2), 16–28.

Tversky, A. & Kahneman, D. 1974. Judgment under Uncertainty: Heuristics and Biases. *Science (New York, N.Y.)*, 185 (4157), 1124–1131.

Ulrich, W. 1983. Critical heuristics of social planning: A new approach to practical philosophy.

Ulrich, W., 1996. *A primer to critical systems heuristics for action researchers*. Centre for Systems Studies Hull.

Van Mierlo, B. & Beers, P.J. 2020. Understanding and governing learning in sustainability transitions: A review. *Environmental Innovation and Societal Transitions*, 34, 255–269.

Westberg, L. & Polk, M. 2016. The role of learning in transdisciplinary research: moving from a normative concept to an analytical tool through a practice-based approach. *Sustainability Science*, 11 (3), 385–397.

Wiek, A.& Iwaniec, D. 2014. Quality criteria for visions and visioning in sustainability science. *Sustainability Science*, 9 (4), 497–512.

Wilson, E.O. 1998. *Consilience: The unity of knowledge*. London: Abacus.

Zurbriggen, C. & Lago, M.G. 2014. Innovación y co-creación: Nuevos desafíos para las políticas públicas. *Revista de Gestión Pública*, 3 (2), 329–361.

Zurbriggen, C., González Lago, M., Baethgen, W., Mazzeo, N., & Sierra, M. 2020. Experimentation in Public Policies: The Uruguayan Soils Conservation Plans. *Iberoamericana – Nordic Journal of Latin American and Caribbean Studies*.

The Future of the Past:
Paths towards Participatory Governance for Cultural Heritage – García (eds)
© 2021 Taylor & Francis Group, London, ISBN 978-1-032-02129-4

Citizen integration in project prioritization methodologies

J. López
Lund University, Lund, Sweden

D. Pulla & J. Carvallo
Universidad de Cuenca, Cuenca, Ecuador

ABSTRACT: Expert-led multicriteria methodologies might have significant improvements in terms of relevance and effectiveness when citizen participation is integrated. This inclusion is especially influential regarding urban revitalization projects (URPs) prioritization, which directly affects the conservation of cultural heritage. The objective is to apply an existing multicriteria methodology for URPs' prioritization in Sígsig, Ecuador and extend it integrating citizen participation. The final methodology uses multiple processes, such as FDM (Fuzzy Delphi Method), ISM (Interpretative Structural Modeling), the integrated phase of CV (Cumulative Voting), and ANP (Analytic Network Process). Results show how citizen integration in these methodologies inherently embeds cultural heritage in the final URP. Conclusions address the importance to integrate participative tools as a way to legitimize urban decision-making processes, along with identifying, strengthening, and preserving intangible heritage, such as traditional craftsmanship, tangible heritage, and archaeological remains.

1 INTRODUCTION

1.1 *Sígsig: National patrimony town, in need for revitalization*

Located in southeast Azuay province in the middle of the Ecuadorian Andes, Sígsig town is privileged to have one of the oldest archaeological sites of Latin America, a Historic Center declared as National patrimony, and an active social dynamic around traditional hat hand-making. However, this town fights migration, caused by low economic opportunities.

A proper urban revitalization project (URP) is needed. Nonetheless, urban project selection in developing towns such as Sígsig are in many cases, unclear. Several times, this selection becomes an instrument for political campaigning, or may even respond to private economic interests.

It's required a well-founded process for urban project selection to contemplate the variety of factors regarding this type of decision-making. "Multiple criteria decision-making (MCDM) is a subdiscipline of operations research that considers numerous criteria in decision-making environments. Be it in our daily lives or in professional engineering settings, there are typically multiple conflicting criteria that need to be evaluated in making decisions." (Chang 2013).

1.2 *Setting a proper methodology*

The goal is to prioritize URPs using a multicriteria methodology which integrates the population in the decision-making process. The study seeks to include not only academics, professionals and community leaders, but the entire community who work and reside in Sígsig. According to Chang, a MCDM methodology results appropriate for this case. It's intended to find a methodology which ponders the case study's positive and negative aspects.

Benefits, opportunities, costs and risks (BOCR) is a concept from the Analytic Network Process, (ANP), proposed by Saaty. It's a simple tool for decision-making, which solves complex multi-attribute problems. BOCR-based multicriteria methodologies create a framework with the BOCR merits. This type of analysis assimilates positive and negative aspects, as well as present and future ones (Wijnmalen 2007).

After analysis and discussion of their content and applicability, 3 potential MCDM methodologies were chosen. A group of parameters were set to evaluate the methodologies, according to the case study and diverse literature references, such as Saaty and Ergu (2015).

These parameters scored each potential methodology, and the highest added score was chosen. Consequently, "An integrated decision-making model for district revitalization and regeneration project selection" by Wang, Lee, Peng & Wu (2013) becomes the framework for the present study.

Wang's methodology develops in three phases according to the four BOCR merits: The first one consists on gathering, categorizing, and prioritizing urban revitalization criteria. For further analysis, the Fuzzy Delphi Method (FDM) is used, pondering the opinion of experts on how to achieve revitalization.

DOI 10.1201/9781003182016-15

On the second phase, the prioritized criteria are evaluated through the relationship between them. A second group of experts establish if there is or isn't a relationship between the criteria using the Interpretative Structural Model (ISM). Criteria without relationship to any other one must be eliminated from further actions.

The third phase is to prioritize an URP, using the criteria from FDM and ISM through ANP. This process creates a group of matrices involving potential URPs and evaluates them according to every criterion and merit. The result is a ranking of the most suitable revitalization urban project.

However, during the MCDM methods analysis, none of them involved the case study population. Most of them were expert-led methods. In Wang's methodology, the case study population has null participation, and the potential URPs are proposed without description of how were they chosen nor by whom. Adjustments must be made to Wang's methodology, adapting it to the Sígsig context.

Resident participation is any act of engaging in project proposal, establishment of project, and implementation by residents who have a direct or indirect interest on whether a certain project is carried out or not. If project planning and implementation can secure rationality and democracy by engaging residents in the processes, then they can not only heighten the sense of community among residents but also contribute to the revitalization of local autonomy. (Woong-Kyoo 2002).

Thereby, the potential projects will come from a deeper diagnosis of Sígsig. An additional phase will be added where the population choose the potential URPs based on their needs, context, and experiences through a participative method.

1.3 *Participative phase*

Policy makers face a difficult dilemma. On one hand, scientific experience is needed, but not enough condition to make pertinent decisions.

Without considering values and public preferences, decisions cannot be legitimized. On the other hand, public perceptions are partially based on biases, anecdotal evidence, and false assumptions about the impacts of human actions (Janse 2007).

This point of view is confirmed by an important reference regarding resident participation in urban renewal, by Yan Hong (2018).

The main advantage of resident-led participation is their great desire to contribute. The main problem is each individual will be strongly determined to have their personal interests and rights guaranteed. Residents should also fulfill the role of improving their environment on a community level. Hence, a combination of administrative-led, expert-led, and resident-led methods is recommended to raise the effectiveness of participation (Hong 2018).

Therefore, several levels and participation methods were pondered to achieve a combined method. According to Hong, there are 8 participation levels; 8th and 7th

(resident control and delegated power) are not related to our study case population reality. However, Sígsig population is compatible and a better fit to the 6th and 5th level (Alliance and Conciliation). The smallest levels; 4th to 1st lay on symbolic or null participation levels.

Within alliance and conciliation level of participation, several communication channels can be found. Sígsig as a town with limited access to Internet connection and technology strings those channels to physical ones.

In addition, it's important to know the priority difference between alternatives. Rinkeviès and Torkar state those particular features can be achieved through Cumulative Voting (CV); a method where Sígsig's population can effectively participate in the decision-making process.

Therefore, the population will select the potential URP that best suit them for further analysis based on their experiences and needs.

2 OBJECTIVE

The present study's purpose can be split in two key points. The first one is to apply a BOCR-based multicriteria methodology to prioritize an urban revitalization project (URP).

The second one is to extend this methodology in order to integrate the case study's population in the decision-making process. This is necessary to generate a tool for heritage features' recognition and development within urban-scaled projects.

Hence, the study contribution is to apply an existing methodology in an Ecuadorian social environment to achieve a justified prioritization.

The necessity to adapt the existing method for integrating population aims to make it more appropriate for our social context.

3 METHODOLOGY

3.1 *Diagnosis review*

To reach the best comprehension level, a context overview is needed. All the information was obtained from the Sígsig's Territorial Plan by Coronel (2015). The diagnosis was developed in 4 dimensions:

Demographic: Sígsig's county population is 26,910 people. 47,58% of the population is between 0–19 years old, 42,36% is between 20–64 years and 10,06% of the population is 65 or older. The principal migration motive is job seeking, with 90,94% of the cases.

Sociocultural: Cultural intangible heritage such as hand hat-making, wood carving and sports are strongly attached to resident's daily lives. In addition, cultural tangible heritage such as the Historic Town Center is a relevant town scenario.

Economic: Economy develops around agriculture, cattle raising, and craftsmanship. However, none of these activities are fully developed nor technical. There has been an increasing touristic activity, around its

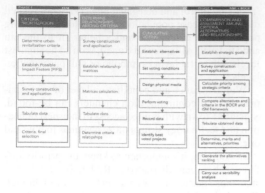

Figure 1. Integrated proposed methodology.

crafts culture (Panama hats), its natural, built and archaeological heritage (Playas de Zhingate Riverside, Historic Center and Chobshi Cave).

Strategic planning: A planning document exists for every project for the 2014–2019 administration. Sígsig is not included in the Ecuador's strategic projects agenda. Also, some problems are low training in machinery and technological investment. Social actors don't participate in the planification nor decision-making processes.

3.2 Integrated methodology

A proposed methodology's synthesis, integrating the new CV participative phase is shown on Figure 1.

3.2.1 Fuzzy Delphi Method (FDM)

FDM is a mathematical tool allows cumulative distributed frequency to generate forecasts from expert opinions in diffuse numbers. It uses the intersection of maximum and minimum ranges to find the results. In this case study, it is necessary to set Possible Impact Factors (PIFs).

These criteria will evaluate the urban projects and prioritize them. First, the criteria must be defined. Since there are many urban revitalization criteria, they should be analyzed and prioritized.

Through exhausting literature review directed to urban revitalization processes, a group of 24 PIFs was gathered: 6 per each BOCR merit.

The PIFs for Benefits are: 1) Strategic regional location; 2) Natural context in good conservation state; 3) Walkable distance in the urban center; 4) Social cohesion among citizens; 5) Identity of construction techniques; and 6) Ideology and local pride.

For Opportunities: 7) Participative design activities; 8) Job opportunities generations; 9) Sustainable mobility systems, 10) Economically-sustainable project; 11) Exportation items' production; and 12) Generation of touristic routes.

For Costs: 13) Importation of goods from cities; 14) Reinvestment for degraded spaces; 15) Municipal works' maintenance, 16) Re-adaptation of buildings; 17) Transport infrastructure's provision, and 18) Conservation of natural and cultural heritage.

Table 1. Prioritized PIFs for FDM method

Merit	PIFs	Fi1			Fi2			Xi*	Limit Value
		C_{i1}	M_{i1}	D_{i1}	C_{i2}	M_{i2}	D_{i2}		
B	1) SRL	2.08	4.16	7.75	8.25	9.16	9.58	8.00	
	2) NCS	5.50	7.33	8.16	8.25	9.16	9.58	8.20	
	3) WDC	1.87	6.00	9.37	7.50	9.16	9.58	9.00	
	4) SCC	7.18	8.12	9.37	9.25	9.50	9.75	9.28	8.014
	5) ICT	4.25	5.50	6.75	9.06	9.37	9.68	7.00	
	6) ILP	3.50	6.00	7.75	7.50	9.16	9.58	7.66	
O	7) PDA	3.50	6.00	8.50	8.25	9.16	9.58	8.33	
	8) JOG	7.25	8.50	9.37	9.25	9.50	9.75	9.28	
	9) SMS	2.25	9.16	9.58	7.50	9.16	9.58	9.16	
	10) EPG	2.25	6.00	9.37	7.50	9.16	9.58	9.00	8.633
	11) EIP	0.62	4.00	7.75	4.25	8.00	9.37	7.00	
	12) GTR	5.25	6.50	8.50	7.83	8.66	9.37	8.00	
C	13) IGC	0.41	0.83	7.75	4.25	9.16	9.37	7.00	
	14) RDS	1.50	6.00	8.75	4.25	9.16	9.58	8.00	
	15) MMI	3.25	4.50	6.50	7.50	9.16	9.58	7.00	
	16) RAB	1.50	4.00	5.75	7.50	8.00	8.75	6.33	7.033
	17) TIP	6.25	7.33	8.16	9.06	9.37	9.68	8.33	
	18) CCH	5.83	6.00	8.75	9.06	9.37	9.68	9.00	
R	19) PAM	3.25	4.50	9.37	8.25	9.16	9.68	9.00	
	20) DNC	5.50	7.50	8.75	9.25	9.50	9.75	9.16	
	21) PCO	0.62	2.00	8.12	4.25	9.16	9.58	7.00	
	22) LIT	5.25	6.50	7.75	8.25	9.16	9.58	8.00	8.003
	23) NEG	1.50	4.00	7.75	8.25	9.16	9.58	8.00	
	24) PNI	6.25	7.50	8.75	9.50	9.37	9.68	9.00	

For Risks: 19) Progressive abandonment; 20) Natural context degradation; 21) Private capitals outflow; 22) Loss of identity and traditions; 23) Null economic growth; and 24) Planning regulations' negative impact.

These PIFs will be evaluated by a group of 5 experts in architecture and urbanism; spatial analysis and urban studies; social geography; urban planning, and economic and socio-territorial demography.

Experts evaluated this 24 PIFs in a survey where they must indicate the importance ideal value (0–10) of each PIF. They must also indicate the range of importance this PIF should have according to their expertise.

Finally, they must indicate their level of expertise in the area of each PIF. Two mathematical function are constructed, based on the frequency of minimum and maximum data per each PIF. Both functions intercept at a certain point.

Then, extracting the slope of both functions, the cross point is found. In order to a PIF to be accepted, the cross-point value must be equal or higher than a limit value.

The limit value comes from the average value of all 5 experts' ideal importance, which was asked in the survey. This process was done to all the 24 PIFs. The prioritized criteria are shaded in gray in Table 1.

3.2.2 Interpretative Structural Modeling (ISM)

The methodology needs to ponder the influence of a variable over other. The prioritized criteria must now

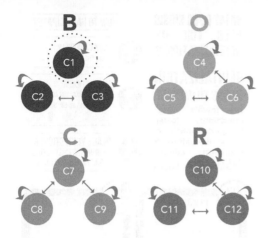

Figure 2. Criteria relationships for each BOCR merit.

be analyzed to identify their relations among them in their BOCR frame (Wang 2013).

This procedure is done through a binary matrix-based survey. One matrix per each BOCR merit is proposed, where their criteria is compared assigning a 1 value when there is a direct relationship, or 0 if there isn't.

Five experts in Architecture; Mobility; Psychology, behavior and perception; Architectural design and heritage preservation analyzed if there is a direct relationship among the prioritized criteria. Criterion which is not related to the rest of its merit criteria will be eliminated from further analysis.

Surveys were performed, and a summary matrix is constructed. The matrix synthesizes the 5 experts' opinions using its mode. There are 4 result matrices per each BOCR merit. The result matrix must be added its identity matrix, and its result will be exposed to its $k+1$ power.

The resulting matrix contains the processed relationships. For a better understanding, Figure 2 is shown. C1: "Natural context in good conservation state" from the Benefits merit is not related to any other Benefits criteria. Therefore, it's eliminated from further analysis.

3.2.3 Cumulative voting (CV)

CV is a polling method which prioritize among various options with multiple winners. It's focused on the whole public sphere through participant's self initiative. Dot-voting is a physic alternative of CV to recognize tendencies in multitudinous decision-making processes. It's used for reaching consensus when stakeholders have high disagreement risk. This process works as a base for deeper prioritization analysis, such as the analytic hierarchy process (AHP) (Segar 2015).

Step 1: Gather potential options through the Diagnosis information: Urban revitalization addresses the dynamics of economic transformation through land use intervention. Therefore, the relevant land uses are which represent the town's public space, and which

may directly affect their competitive development. Hence, it's mandatory to analyze the land uses' classification within the studied environment. Ecuador's relevant planning documents were considered to refine the pertinent land uses for the projects. The five relevant land uses are: multiple, heritage area, industrial, equipment, and service and commerce activities.

Potential project generation: To generate urban project captions, the relevant land uses are analyzed, along with the Diagnosis review information. Potentialities, problems and needs were outlined and related with the land uses to generate urban projects for further resident prioritization.

O1: Managerial Groups aims to re-establish links and discussion spaces between the population and the administrative authorities. O2: Trade Zones focuses on fostering local production commerce, such as crafts and food production. O3: Industrial Zones aims to boost agricultural production to reach competitive levels.

O4: Cultural Scenarios aims to revalue their endangered archaeological and immaterial cultural heritage. O5: Crafts Training focuses on specializing and improving traditional knowledge to generate economic development. O6: Public space intervention intends to improve the built environment to promote urban revitalization. O7: Sport Specialization fosters sport equipment construction as a way to revitalize people's way of living. Finally, O8: Touristic spots aims to promote Sígsig's cultural heritage to reach economic development.

Step 2: Create a caption for the voting purpose in one single sentence. The caption and definition are shown in Figure 3.

Step 3: Identify the finite prospective demographic sample for the voting: According to the equation for statistical demographic sampling proposed by Aguilar-Barojas (2005), 199.15 participants are needed, which rounds up to 200 voters. The Sígsig urban population of 11,170 inhabitants, an absolute accuracy of 3% and a significance level of 95% were used. In addition, 5 age ranges where proposed: 15–25; 26–35; 36–50; 51 -65, +65, according to Sígsig's demographic distribution.

Step 4: Identify the voting's optimum environment: According to visits and resident interviews, the biggest people affluence happens on weekends, around town's important landmarks. The chosen location for the voting was the Town Market's outdoors on Sunday around 10:00am.

Step 5: Create the physical medium for vote placing: In order to withstand around 200 votes, be easy to carry and be visible from a distance, a cardboard tabloid measuring 1.85mx0.9m was prepared. A vertical alternative is shown on Figure 3.

Step 6: Conduct the voting in base of the established guidelines: Residents approached the tabloid. Then, a brief explanation about the objective, the projects and how to vote was provided. Finally, each voter took a single-color sticker according to their age range and placed it in the URP that better fit their opinion.

Step 7: Re-establish voting in one-hour time periods: After an hour passed, the voting tabloid was

Table 2. Cumulative voting results

Age ranges	15–25	26–35	36–50	51–65	+65	Total	Rank.
O1: Manag. groups	2	12	14	6	8	42	3rd
O2: Trade Zones	2	10	11	6	5	34	4th
O3: Industrial zones	0	4	4	1	0	9	7th
O4: Cult. scenarios	2	0	4	0	0	6	8th
O5: Crafts Training	8	21	12	14	5	60	1st(e)
O6: Public space int.	5	7	10	2	3	27	5th
O7: Sport specializ.	8	2	1	0	1	12	6th
O8: Tour. spots	18	15	16	6	5	60	1st(e)
Total	45	71	72	35	27	250	
(%)	18%	28.4%	28.8%	14%	10.8%	100%	

re-established, registering the included votes and switching the option's order. Voting is continued in the same location.

Step 8: Repeat Step 7 until reaching the needed votes: After another one-hour and a half period and 2 voting rounds, the voting was finished. An additional 50 participations were accomplished.

Step 9: Record the obtained data: Photographic registration was used to record the votes in an effective way between one-hour periods.

The data registered by projects and by age range are detailed in Table 2. The age range data is consistent with the Sígsig's demographic distribution, which conveys additional significance to the sample.

The four most voted projects will continue the further prioritization analysis: ANP. The most voted projects are the following; Crafts training and Touristic spots with 60 votes each; Managerial groups with 42 votes; and Trade zones with 34 votes.

3.2.4 Analytic Network Process (ANP) with BOCR

ANP with BOCR is a multi-attribute framework to ponder simultaneously positive and negative impacts of a single problem. The evaluation model consists of a control hierarchy and BOCR sub-networks. The hierarchy establishes the merits' values. The results from FDM, ISM and CV are used to build four sub-networks, and a final ranking to prioritize the most efficient URP can be generated.

First, a general objective and three strategic goals should be stated. The Diagnosis phase is especially relevant to formulate them.

O1: To achieve urban revitalization through a justified urban project prioritization; SG1: "To boost local production to achieve economic development"; SG2: "To foster production trading in Sígsig urban county"; SG3: "To relieve the migration's effect on demographic growth".

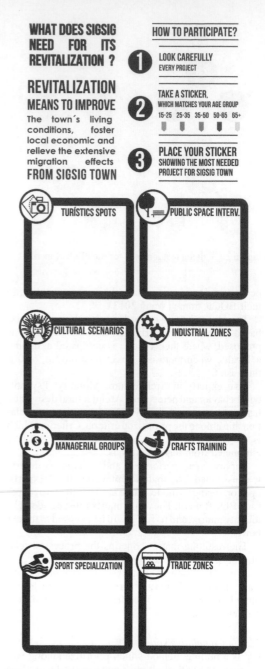

Figure 3. Vertical tabloid for Dot-voting.

The ANP process is achieved with an operation called pairwise comparison, which is every comparing process between pairs of entities to judge which one is preferred, has a higher determined quantified property, or if the two entities are identical or not. An extensive survey with multiple comparison matrices must be developed.

Five experts directly related to Sígsig's reality were invited. These experts are strategic locals who have worked in Sígsig's administration and who know about the history, society, economy, development demands,

Table 3. Final ranking of prioritized URPs

Alternatives	Additive		Prob. Additive		Subtractive	
	Prior.	Rank.	Prior.	Rank.	Prior.	Rank.
A1: Crafts training	0.27073	2	0.50937	2	0.02826	2
A2: Touristic Spots	0.28205	1	0.52413	1	0.04302	1
A3: Managerial groups	0.22538	3	0.47109	3	−0.01002	3
A4: Trade Zones	0.22184	4	0.45763	4	−0.02348	4

and territorial dynamics of Sígsig town and county. Their expertise fields are strategic planning, urban control, municipal decision-making, and collective groups leadership.

In the model's first part, experts were asked to evaluate the three strategic criteria's priority. The data matrices from all the survey should be checked with a consistency index and ratio.

The integrated priorities of the strategic criteria are: SG1= 0.1326; SG2=0.3347; SG3=0.5326. This means SG3 is the most important strategic goal to achieve urban revitalization in Sígsig.

For the model's second part, the alternatives are evaluated under each BOCR merit and its sub-network. One matrix is constructed per case and expert. At the end, results for each criterion are generated. These go into a synthesis matrix where lay all the experts' values. The values are averaged to obtain a final value. Furthermore, each project alternative should be compared between themselves in the framework of the 11 ISM criteria.

An unweighted matrix per BOCR merit is constituted and needs to be processed to get a limit matrix. The mathematical process to get a limit matrix can be simplified using a software called SuperDecisions. This is an open software based on AHP and ANP for decision-making. Inside the software, the framework must be constructed, including the control hierarchy and the four sub-networks with the FDM criteria, their ISM relationships, and the four alternatives from CV. The limit matrix conveys the priority of each alternative regarding each BOCR sub-network.

The experts assessed the priority of the four merits according to the strategic criteria. The final priorities of each BOCR merits are called normalized priority, with the following results: B=0.23416, O=0.28472; C=0.25046 and R=0.23065.

To conclude the ANP process and get the final alternative ranking, the additive, probabilistic additive, and subtractive equations can be used, as stated by Wang. The final results are shown in Table 3.

Using the 3 synthesis ranking methods, the prioritized projects have the same order. The project A2 has the biggest priority, followed by A1, A3, and A4.

A final ranking of urban revitalization criteria was also developed. The most important criteria were C2, Walkable distance in the urban center, followed by C9,

Conservation of natural and cultural heritage, and C6, Economically-sustainable project. The prioritized criteria reveal development key points for the prioritized project.

4 RESULTS

4.1 Dot-voting Method as a tool for participative governance

In order to generate the voting potential projects, the Sociocultural review is a key source of potentialities and necessities. As a result, 4 of the 8 options were derived from Sígsig's cultural richness. After Dot-voting, the results showed an unusual tie between Crafts training and Touristic spots, each with 60 votes from the 250 total votes. These results are especially significant for the decision about the Sígsig final URP. This whole process embeds cultural heritage and its dimensions in the potential URPs.

The method acts as an experimental resident survey for further analysis and has its own limitations. It can be deepened to achieve a better resident integration to ponder and propose justified, communal projects, and to extend the number of participations.

4.2 Assessing final results from the methodology

Considering the additive method in the final URPs ranking, a difference of 0.01132 points between A1 and A2 is observed. This is a small gap, compared to the third option's difference: more than four times bigger (0.04492) from the second one. A sensitivity analysis is needed to verify the ranking' robustness, increasing or decreasing merit's priorities with a trial and error method until the first prioritized project changes. In the 8 possible cases, only 2 of them change the final ranking. Both of the cases (when Opportunities merit decreases from 0.28472 to 0.190003, and Risks increases from 0.23065 to 0.29932) benefit A1 alternative: Crafts training. Both changes are within reasonable and possible priorities' shift, according to the expert's opinion.

The last two considerations mean A2 alone as the best alternative for urban revitalization for Sígsig town is a strong solution, but the importance and priority that A1 has to complement A2 needs to be considered as a more complete and potentially successful direction for urban renewal. Even though A3 and A4 have a lower priority, both can work towards urban revitalization.

The proposal is to build up a business network which contribute to the strategic goals. Henceforth, the project with the biggest priority builds up the foundations for urban revitalization, and the next projects cover specific issues towards the same goal.

4.3 Artisan tourism: a concept centered on cultural heritage

The final ranking for URPs shows the best option to be A2: Touristic spots with 0.28205 and A1: Crafts training with 0.27073. The small gap between the two first options confirms the unusual tie in the Dot-voting method between the same alternatives.

Both of them are almost equally important to achieve urban renewal in Sígsig, according to the expert's panel and the population.

This analysis raises the option to conform a single revitalization project with the options which had the biggest priority in the prioritization process. This project would act in different problematics simultaneously (Touristic spots with Crafts training), developing a concept for Sígsig objective, validated by both the experts and the population.

The artisan tourism development in Sígsig town as a revitalization tool pledges to change realities, such as progressive abandonment, heritage neglect, and lack of economic development. This concept holds at its core the promotion of the town's cultural heritage. On one hand, immaterial cultural heritage with traditional craftsmanship. On the other hand, all material cultural heritage: such as Playas de Zhingate riverside as a natural zone, the Historic Center as built environment, and the abundant archaeological remains, such as Chobshi Cave representing both cultural and natural heritage.

This remarks the importance of considering the town's cultural patrimony to help solving their deepest problems.

5 CONCLUSIONS

Sígsig town presents a need of revitalization, which inspire the application of an existing multicriteria methodology to prioritize URPs. Citizen integration in decision-making processes is an important contribution to expert-led methods.

In this case, the Sociocultural Diagnosis review was especially relevant. Therefore, traditional craftsmanship, tangible heritage, and archaeological remains were identified, strengthened, and preserved through the proposed projects.

The integration of a citizen participation tool such as Dot-voting has proved effective as a way to synthetize population's opinion regarding URPs relevance. Since inhabitants are involved in the selection, a specific derived architectural project is more likely to have success. This is a clear advantage in comparison to solely expert-led decision-making.

In addition, it's a step forward to democratize and legitimate urban decision-making in the Ecuadorian Andes. The experts' panel and the population agreed in the decision of the best way to achieve urban revitalization in Sígsig.

The final URPs ranking is a useful tool to understand the priority of each project and their relation with each other to decide the generation of potential compound projects. In this case, this option might be more effective and relevant. A combination of the two higher-priority projects is proposed to formulate a new and integral URP, starting from the ranking analysis.

The two projects are complementary with each other, and are heritage-centered in their own way. Artisan tourism is a concept based on cultural heritage, which has been conceived as the spinal column that fosters economic development for Sígsig town.

Finally, citizen integration in project prioritization methodologies becomes an excellent path to follow towards territorial participative governance; where local problems are diagnosed, authorities work for the greater good, stakeholders are compromised with their communities, heritage is appreciated and boosted, and all the voices are valued.

ACKNOWLEDGMENTS

We thank Arch. PhD. Fernanda Aguirre for the support to structure this article.

REFERENCES

Aguilar-Barojas, S. 2005. Fórmulas para el cálculo de la muestra en investigaciones. Villahermosa, México. 338 páginas.

Chang, C. 2013. Multiple Criteria Decision-Making Theory, Methods, and Applications in Engineering. Chang Gung University, Taiwan. https://www.hindawi.com/journals/mpe/si/970806/cfp/.

Coronel, A., Vásquez, G., Saquisilí, N. 2015. Actualización PDYOT. Geoliderar, GAD Municipal del Cantón Sígsig, Azuay, Ecuador.

Hong, Y. 2018. Resident participation in urban renewal. Frontiers of Architectural Research. https://doi.org/10.1016/j.foar.2018.01.001.

Janse, G., & Konijnendijk, C. C. 2007. Communication between science, policy and citizens in public participation in urban forestry. Urban Forestry & Urban Greening, 6(1), 23–40. https://doi.org/10.1016/j.ufug.2006.09.005.

Saaty, T., Ergu, D. 2015. When is a Decision-Making Method Trustworthy? Criteria for Evaluating Multi-Criteria Decision-Making Methods.

Segar, A. 2015. The Power of Participation: Creating Conferences That Deliver Learning, Connection, Engagement, and Action. 322 páginas. ISBN: 9781511555982.

Wang, W.-M., Lee, A. H. I., Peng, L.-P., & Wu, Z.-L. 2013. An integrated decision-making model for district revitalization and regeneration project selection. Decision Support Systems, 54(2), 1092–1103. https://doi.org/10.1016/j.dss.2012.10.035.

Wijnmalen, D. 2007. Analysis of benefits, opportunities, costs, and risks (BOCR) with the AHP–ANP: A critical validation, Mathematical and Computer Modelling, Volume 46, Issues 7–8, ISSN 0895–7177. https://doi.org/10.1016/j.mcm.2007.03.020.

Woong-Kyoo, B., 2002. Housing improvement and citizen participation for the recovery of urban community. Rev. Archit. Build. Sci. 46(11),44–49.

The Future of the Past:
Paths towards Participatory Governance for Cultural Heritage – García et al (eds)
© 2021 Taylor & Francis Group, London, ISBN 978-1-032-02129-4

Author index